Earthmate

Ali Faridi Masouleh

Earthmate

Rethink, Revive, Sustain: A Bold Blueprint for a Greener Tomorrow

Ali Faridi Masouleh
London, UK

ISBN 978-3-031-82555-2   ISBN 978-3-031-82556-9 (eBook)
https://doi.org/10.1007/978-3-031-82556-9

© The Editor(s) (if applicable) and The Author(s), under exclusive license to Springer Nature Switzerland AG 2025

This work is subject to copyright. All rights are solely and exclusively licensed by the Publisher, whether the whole or part of the material is concerned, specifically the rights of translation, reprinting, reuse of illustrations, recitation, broadcasting, reproduction on microfilms or in any other physical way, and transmission or information storage and retrieval, electronic adaptation, computer software, or by similar or dissimilar methodology now known or hereafter developed.
The use of general descriptive names, registered names, trademarks, service marks, etc. in this publication does not imply, even in the absence of a specific statement, that such names are exempt from the relevant protective laws and regulations and therefore free for general use.
The publisher, the authors and the editors are safe to assume that the advice and information in this book are believed to be true and accurate at the date of publication. Neither the publisher nor the authors or the editors give a warranty, expressed or implied, with respect to the material contained herein or for any errors or omissions that may have been made. The publisher remains neutral with regard to jurisdictional claims in published maps and institutional affiliations.

This Springer imprint is published by the registered company Springer Nature Switzerland AG
The registered company address is: Gewerbestrasse 11, 6330 Cham, Switzerland

If disposing of this product, please recycle the paper.

To the green-hearted souls who dare to dream beyond themselves, who nurture life with their compassion, and who walk the path of true enlightenment. This book is a tribute to your boundless courage, your enduring hope, and your unwavering commitment to a world that thrives in harmony and grace.

Preface

Congratulations on choosing this book. This decision reflects the depth of your humanity, as in today's overly profit-driven, illogical, and unhealthy world, you have chosen to explore a book that aims to address interests beyond oneself.

From childhood, I have been deeply concerned about our Earth and humanity. Questions haunted me: Why don't things proceed logically? Or perhaps, why don't they align with the logic of someone who cares about the Earth and its creatures? I was preoccupied with issues that, I believe, few others have deeply engaged with. Historical conflicts seem to linger, perpetuating misunderstandings, suspicions, and misinterpretations that worsen our global crises. As I write this, the world faces ongoing tensions in regions like Ukraine and Russia, Palestine, Yemen, Lebanon, Syria, and Israel alongside disputes involving the Koreas, the US, Taiwan and China and more. In these conflicts, nature and vulnerable creatures are victims of these conditions and often become collateral damage.

I have always pondered the possibility of a united, conscious, and robust global effort to protect our shared home—the Earth. Though it may not solve all problems, such an alliance could address a significant portion of the greed-driven and short-sighted issues plaguing humanity.

Some may consider my perspective idealistic, romantic, or even naïve. However, I believe idealism—when paired with realism and practical strategies—can serve as an inspiring force. Throughout history, groundbreaking changes often began with seemingly unreachable ideas. This book combines idealism with science-based facts and actionable solutions, offering a balanced perspective that aims to inspire and inform.

My approach is not romantic but rooted in humanism. I aim to address the deep concerns of our time, not merely as an observer but as someone who believes in the potential for positive change. As you delve into this book, you will find philosophical and ethical reflections that strive to comprehend the root causes of our challenges rather than simply highlighting them.

This work is not an attempt to solve the complex problems tied to historical, social, economic, and political contexts. Rather, its purpose is to raise questions and provoke thoughtful engagement. The depth of our interaction with such works determines their impact, and I hope this book inspires meaningful reflection and action.

My fascination with these topics began in childhood and shaped my academic journey in geography. During my university years (2011–2017), I began compiling scattered notes that eventually evolved into this book. Between then and the signing of the book contract in 2023, the writing process slowed at times but continued steadily. After the signing of the book contract in 2023 until today, the content underwent numerous revisions and restructuring until it became the book you now hold.

One of the most significant challenges was deciding to write my first book in a language other than my native Persian. Given the low readership among Persian speakers and the limited reception of my previous translations, I realized that reaching a broader audience required stepping out of my comfort zone.

This book is designed for a wide range of readers—scientists, artists, cultural enthusiasts, and the general public. While initially planned as a specialized academic work, I decided to simplify and make it more accessible after signing the publishing contract. Writing in an engaging, straightforward style allows this book to reach more readers and foster greater awareness. Issues like environmental sustainability are global concerns that need to be communicated in an understandable way to drive collective action. This approach aligns with my personality as a writer and reflects my sense of responsibility to educate and inspire a broader audience.

Finally, I wish to express my heartfelt gratitude to Dr. Abbas Saeedi and Springer Nature for their unwavering support and trust in a first-time author. Your encouragement made this journey possible.

London, UK Ali Faridi Masouleh
November 2024

Contents

1	**The Earth**	**1**
	Introduction	1
	The History of the Earth	3
	Hadean Eon (4.6–4.0 Billion Years Ago)	4
	Archean Eon (4.0–2.5 Billion Years Ago)	5
	Proterozoic Eon (2.5 Billion–541 Million Years Ago)	6
	Paleozoic Era (541–252 Million Years Ago)	8
	Mesozoic Era (252–66 Million Years Ago)	9
	Cenozoic Era (66 Million Years Ago—Present)	10
	Engaged Scientific and Intellectual Groups and Majors	11
	Do They Actually Work?	23
	Natural Resources	23
	Climate Patterns	24
	Biodiversity	25
	Human Population	26
	Environmental Impact	27
	Ocean Health	29

Land Use	29
Natural Disasters	30
Air Quality	32
Conclusion	33
References	34

2 Earthmate Philosophy — 37
Introduction — 37
Deus sive Natura — 38
Absolute Idealism — 40
Naturphilosophie — 42
Collectivism Versus Individualism — 43
Organicism Versus Mechanism — 47
Unity Despite Plurality (UDP) — 49
References — 52

3 Earthmate Science — 55
Introduction — 55
Knowledge and Science — 61
  Complex System — 63
  Systems Thinking — 65
  Interdisciplinarity — 66
  Sustainable Development — 68
  Scarcity and Finite Resources — 70
  Circularity or Circular Economy (CE) — 71
  Environmental Fragility — 74
Scientific Actions — 76
  Prevention of Environmental Problems — 76
  Conservation and Restoration of Ecosystems and Biodiversity — 77
  Adapting to and Building Resilience Against Environmental Changes — 78
Conclusion — 79
References — 81

4	**Earthmate Ethics**	85
	Introduction	85
	Respect for Nature	91
	Interdependence	94
	Responsibility and Sustainability	97
	Equity and Justice	99
	Collaboration, Tolerance and Shared Goals	101
	Conclusion	102
	References	103
5	**Earthmate in Action**	105
	Introduction	105
	Section 1: The Role of Individuals, Communities, and Organizations	105
	Section 2: Successful Practices and Initiatives	106
	Section 3: Challenges and Obstacles	106
	The Role of Individuals, Communities, and Organizations	107
	Individuals	107
	Educational and Consultative Institutions	114
	Media and Culture	117
	Profit-Oriented Entities	119
	Non-profit Entities	126
	Governmental Entities	130
	Health and Environment	138
	International Organizations	141
	Special Environment-Oriented Entities	145
	Successful Practices and Initiatives	148
	Individuals	148
	Educational and Consultative Institutions	154
	Media and Culture	157
	Profit-Oriented Entities	159
	Non-profit Entities	164
	Governmental Entities	168
	Health and Environment	172
	International Organizations	174

Special Environment-Oriented Entities	177
Challenges and Obstacles	179
Individuals	179
Educational and Consultative Institutions	186
Media and Culture	190
Profit-Oriented Entities	192
Non-profit Entities	198
Governmental Entities	202
Health and Environment	206
International Organizations	208
Special Environment-Oriented Entities	212
Conclusion	214
References	216
6 The Future of Earthmate	**219**
The Potential Impact of the Earthmate Principles	219
Innovations and Advancements in the Earthmate Principles	223
The Role of Principles Earthmate in Shaping the Future	225

Abbreviations

CBD	Convention on Biological Diversity
CCS	Carbon Capture and Storage
CE	Circular Economy
CEAP	Circular Economy Action Plan
CO_2	Carbon Dioxide
CSR	Corporate Social Responsibility
DAC	Direct Air Capture
ECMWF	European Centre for Medium-Range Weather Forecasts
EPA	Environmental Protection Agency
ESS	Earth System Science
ESSP	Earth System Science Partnership
EU	European Union
FAO	Food and Agriculture Organization of the United Nations
GCF	Green Climate Fund
GEC	Global Environmental Change
GECAFS	Global Environmental Change and Food Systems
GEF	Global Environment Facility
GFDRR	Global Fund for Disaster Risk Reduction
GHG	Greenhouse Gas
GIS	Geographic Information Systems

GMSL	Global Mean Sea Level
HDEs	Halo Disturbance Effects
HEAL	Health and Environment Alliance
ICSU	International Council for Science
IEA	International Energy Agency
IGBP	International Geosphere-Biosphere Programme
INECE	International Network for Environmental Compliance and Enforcement
IPAT	I = PAT or I = P × A × T, (environmental) Impact, Population, Affluence, Technology
IPCC	Intergovernmental Panel on Climate Change
IRENA	International Renewable Energy Agency
IUCN	International Union for Conservation of Nature
IWRM	Integrated Water Resource Management
LCA	Life Cycle Assessments
MOOCs	Massive Open Online Courses
NASA	National Aeronautics and Space Administration
NbS	Nature-based Solutions
NGOs	Non-Governmental Organizations
NOAA	National Oceanic and Atmospheric Administration
NOx	Nitrogen Oxides
O_2	Oxygen
PES	Payment for Ecosystem Services
PM10	Particulate Matter
PPPs	Public-Private Partnerships
R&D	Research and Development
REN21	Renewable Energy Policy Network for the 21st Century
SDGs	Sustainable Development Goals
SEPA	Swedish Environmental Protection Agency
SO_2	Sulfur Dioxide
SOx	Sulphur Oxides
UDP	Unity Despite Plurality
UN	United Nations
UNCED	United Nations Conference on Environment and Development
UNEP	United Nations Environment Programme
UNESCO	United Nations Educational, Scientific and Cultural Organization
UNFCCC	United Nations Framework Convention on Climate Change
UNGA	United Nations General Assembly

USEPA	United States Environmental Protection Agency
VOCs	Volatile Organic Compounds
WEF	World Economic Forum
WHO	World Health Organization
WMO	World Meteorological Organization
WTO	World Tourism Organization
WWF	World Wildlife Fund

1

The Earth

Introduction

"You don't know what you've got until it's gone". An imaginary situation in which the Earth is no longer habitable is a bleak and terrifying scenario that may get real. Without a habitable planet, humanity and all other life on Earth would be forced to confront an existential crisis of unparalleled magnitude.

Imagine a world where Earth has become a wasteland, a stark contrast to the vibrant blue and green planet it once was. The skies, once clear and filled with the songs of birds, are now perpetually shrouded in a thick, gray haze. The air, heavy with toxic pollutants, makes breathing without protective masks impossible.

The oceans, once teeming with life, have turned into vast expanses of acidic, lifeless water. Coastal cities lie abandoned, their streets flooded by rising sea levels. Once majestic forests have been reduced to desolate plains, with only charred stumps remaining as silent witnesses to their former glory.

Temperatures have soared to unbearable levels, with heatwaves sweeping across the continents. Deserts have expanded, encroaching on once fertile lands, leaving behind cracked earth and withered vegetation.

Freshwater sources have dried up, leading to widespread droughts and a desperate scramble for the last remnants of drinkable water.

The once bustling cities are now ghost towns, their skyscrapers standing as hollow shells, slowly crumbling into decay. Nature has reclaimed some areas, but not with the beauty of forests and meadows. Instead, invasive and hardy species, capable of surviving the harsh conditions, dominate the landscape.

Wildlife, if it exists at all, is scarce. Many species have gone extinct, unable to adapt to the rapid changes. The few remaining animals are gaunt and desperate, scavenging for any scraps of food they can find. The balance of ecosystems has been irreparably disrupted, leading to a cascade of ecological collapse.

Humanity's remnants are scattered, living in isolated, heavily fortified enclaves. Technology that once connected the world now serves as a lifeline, providing the basics for survival—clean air, water, and food—within these fortified zones. Communication between enclaves is limited and fraught with danger, as the outside world is filled with hostile environments and desperate scavengers.

In this bleak future, hope is a rare commodity. The dream of finding a new home among the stars is the only thing that keeps the survivors going. Scientists and engineers work tirelessly to develop spacecraft capable of reaching distant, habitable planets, but the journey is fraught with uncertainty. The knowledge that Earth is no longer a sanctuary hangs heavy, a constant reminder of humanity's fragility and the consequences of its actions.

If the Earth were no longer habitable, humans would have to find a new planet to call home, a task that is currently beyond our technological capabilities. Even if we were able to find a suitable planet, the process of traveling there and establishing a new civilization would be fraught with challenges and uncertainties.

The loss of Earth's biodiversity and natural resources would have profound implications for the survival of many species and the functioning of ecosystems. The loss of the Earth's oceans, forests, and other natural environments as a bio-system would be a devastating blow to the planet's natural beauty and cultural heritage.

The collapse of Earth's social, economic, and political systems would also be a major concern. Without a habitable planet to sustain us, the foundations of human civilization would crumble, leaving us without food, water, or shelter. Societal upheaval and conflict would likely ensue, as people compete for the limited resources that remain.

In short, a world without a habitable Earth as a bio-system would be a world of unimaginable suffering, where the very survival of humanity is at stake. It is therefore imperative that we take action now to protect the health and well-being of our planet, ensuring that it remains a viable home for all life for generations to come.

The History of the Earth

Welcome to the captivating world of the Earth's history, a chronicle that spans over 4.5 billion years. From the primordial beginnings of our planet to the rise and fall of ancient civilizations, the history of the Earth is replete with fascination, enigma, and exhilaration.

It is intriguing to note that the Earth once existed as a ball of molten rock, devoid of life or atmosphere. It took billions of years for the first organisms to develop and evolve into the diverse range of flora and fauna we witness today.

Throughout its existence, the Earth has undergone phenomenal transformations, from colossal volcanic eruptions to catastrophic meteor impacts. It has also housed some of the most remarkable and awe-inspiring creatures that have ever existed, such as the mighty dinosaurs and the majestic mammoths.

Nevertheless, the Earth's history is not merely a narrative of geology and biology. It is also a saga of human civilization, from the earliest hunter-gatherer societies to the contemporary technological epoch. Our forebears have left an indelible mark on the planet in innumerable ways, from prehistoric cave art to towering edifices.

Join me on this journey through the annals of time and space, as we traverse the Earth's history and uncover the extraordinary marvels that it holds. From the deepest abysses of the oceans to the towering peaks of the highest mountains, there is always a new and riveting discovery waiting to be made. So, are you ready? Let us embark on this expedition together!

Before beginning of it we need to know about the difference between Era and Eon. Both "Era" and "Eon" are geological time units used to divide the Earth's history into manageable parts. However, there are some differences between the two terms:

- Duration: An era is a subdivision of a geological period, which typically lasts tens to hundreds of millions of years. An eon, on the other hand, is the largest subdivision of geological time, lasting billions of years.
- Scale: Eons are much larger than eras in terms of the amount of time they encompass. There are only four eons in the geological time scale, while there are many eras.
- Events: The boundaries between eras are defined by significant events in Earth's history, such as mass extinctions or major shifts in climate. The boundaries between eons are defined by even more significant events, such as the formation of the Earth or the appearance of complex life.

Hadean Eon (4.6–4.0 Billion Years Ago)

The captivating realm of Hadean Eon takes us back to the earliest days of planet Earth, approximately 4.6 billion years ago. The Hadean Eon was characterized by extreme heat, incessant bombardment by asteroids, and the formation of the planet's first continents and oceans.

The name "Hadean" originates from the Greek deity Hades, the god of the underworld, chosen due to the Hadean Eon's fiery conditions, much like the fires of Hades. During the Hadean Eon, the Earth was in its formative stages, and it was a tumultuous and turbulent place. Massive asteroids and other debris collided with the planet, causing significant impacts and crater formations.

In addition, the Hadean Eon was a period of intense volcanism, as molten magma erupted from the Earth's interior, creating new landmasses and shaping the landscape. Despite the extreme conditions, evidence suggests that life may have already existed on Earth during the Hadean Eon, as microscopic fossils found in rocks that are 3.5 billion years old suggest the existence of ancient bacteria.

The formation of the moon is one of the most significant events of the Hadean Eon. According to the most widely accepted theory, a Mars-sized object collided with Earth, and the debris from the impact eventually coalesced to form the moon.

Moreover, the Hadean Eon is when the Earth's first oceans and continents began to form. The intense volcanism released gases that eventually formed water vapor, which then condensed and fell as rain, filling the planet's basins and creating the first oceans. The first continents were formed through plate tectonics, where the movement of the Earth's crust caused landmasses to collide and merge, eventually forming larger land masses.

In conclusion, the Hadean Eon was a momentous and crucial era in the history of our planet. It was a period of intense geological activity, the formation of the moon, and the emergence of the first oceans and continents. Despite the extreme conditions, life may have already existed on Earth during this time, setting the stage for the evolution of the planet's living organisms in the future.

Archean Eon (4.0–2.5 Billion Years Ago)

The Archean Eon, also known as the "age of the ancient life," is an era in Earth's history that spans from about 4 billion to 2.5 billion years ago. During this time, the planet was a much different place than it is today, with a vastly different atmosphere, climate, and geological features. Exploring the Archean Eon is like taking a trip back in time to a mysterious and fascinating world that holds many secrets and wonders.

One of the most intriguing aspects of the Archean Eon is the emergence of life. While there is still much debate about how and when life originated on Earth, it is believed that the first forms of life appeared during this time. These early life forms were likely simple, single-celled organisms that lived in the oceans and fed on organic matter.

Another fascinating aspect of the Archean Eon is the geological activity that took place. During this time, the Earth's crust was constantly shifting and changing, with frequent volcanic eruptions and earthquakes. This geological activity gave rise to some of the earliest rocks and minerals on the planet, such as granites and gneisses, which can still be found today.

In addition to the geological activity, the Archean Eon also saw the formation of some of the earliest land masses. These early continents were small and fragmented, but they provided a new habitat for life to evolve and adapt to. It is believed that the first land plants appeared during this time, paving the way for the diverse ecosystems that exist on Earth today.

Perhaps the most significant event of the Archean Eon, however, was the formation of the Earth's atmosphere. Prior to this time, the atmosphere was largely composed of carbon dioxide and other gases, but the emergence of photosynthetic organisms began to change that. These organisms, which could convert sunlight into energy, began to release oxygen as a byproduct, gradually changing the composition of the atmosphere and paving the way for more complex forms of life to evolve.

As we explore the Archean Eon and its many wonders, we are reminded of the incredible journey that our planet has been on, and the many mysteries that still remain to be uncovered. From the emergence of life to the formation of the Earth's atmosphere, the Archean Eon holds many clues to our planet's past, and many possibilities for its future.

Proterozoic Eon (2.5 Billion–541 Million Years Ago)

The Proterozoic Eon, spanning from approximately 2.5 billion years ago to 541 million years ago, is a fascinating period in Earth's history that is often overlooked. In this epoch, the planet experienced significant changes that set the stage for the evolution of complex life forms. Allow me to take you on a journey through the Proterozoic Eon, where we will explore some of the most exciting and intriguing aspects of this remarkable period.

First and foremost, it is worth noting that the Proterozoic Eon witnessed the emergence of the first eukaryotic cells. These cells are characterized by the presence of a nucleus and other organelles, which enabled them to carry out complex metabolic processes. This was a crucial step towards the development of multicellular life forms, as eukaryotic cells could form complex structures and perform specialized functions.

In addition to the emergence of eukaryotes, the Proterozoic Eon was also marked by the proliferation of cyanobacteria. These simple, single-celled organisms played a critical role in shaping the Earth's atmosphere, as they were responsible for the first significant oxygenation event. Through photosynthesis, cyanobacteria produced oxygen, which accumulated in the atmosphere and eventually led to the formation of the ozone layer. This layer protected the Earth from harmful UV radiation, paving the way for the evolution of more complex life forms.

One of the most remarkable events in the Proterozoic Eon was the "Great Oxygenation Event," which occurred approximately 2.4 billion years ago. During this period, the oxygen levels in the atmosphere rose dramatically, leading to the extinction of many anaerobic organisms that could not survive in the newly oxygenated environment. This event was a turning point in the history of life on Earth, as it created new ecological niches and paved the way for the emergence of more complex life forms.

Another fascinating aspect of the Proterozoic Eon was the formation of the supercontinent Rodinia. This massive landmass, which existed approximately 1 billion years ago, was the precursor to Pangaea, the supercontinent that formed during the Paleozoic Era. The assembly of Rodinia had a profound impact on the planet's geology, climate, and evolution, as it led to the formation of new mountain ranges and altered ocean currents.

Finally, it is worth noting that the Proterozoic Eon was not without its challenges. During this period, the Earth experienced multiple glaciation events, which had a significant impact on the planet's climate and ecosystems. These ice ages led to the formation of extensive glacial deposits and altered the distribution of life forms around the globe.

In conclusion, the Proterozoic Eon was a critical period in Earth's history that set the stage for the emergence of complex life forms. From the evolution of eukaryotic cells to the oxygenation of the atmosphere, this epoch was marked by significant changes that had a profound impact on the planet and its inhabitants. Whether you are a geologist, biologist, or simply someone who is curious about the history of our planet, the Proterozoic Eon is a fascinating and captivating period that is well worth exploring.

Paleozoic Era (541–252 Million Years Ago)

This is a period of time that spans over 300 million years and is home to some of the most incredible and awe-inspiring creatures that have ever lived. From the first appearance of multicellular life to the rise of the first land animals, the Paleozoic Era is a time of incredible change and evolution.

One of the most remarkable features of the Paleozoic Era is the incredible diversity of life that existed during this time. From the trilobites and brachiopods that dominated the early seas to the amphibians and reptiles that emerged later on, the Paleozoic Era was a time of incredible innovation and experimentation in the animal kingdom.

But it wasn't just the animals that were evolving during this time. The Paleozoic Era was also a time of incredible geological change, with the continents shifting and colliding to form the supercontinent of Pangaea. This in turn had a profound effect on the climate and environment, leading to the formation of vast coal swamps and the eventual rise of the first forests.

Perhaps one of the most fascinating aspects of the Paleozoic Era is the emergence of the first land animals. From the first amphibians that crawled out of the water to the giant insects that roamed the forests, the Paleozoic Era saw the birth of an entirely new kind of life.

But it wasn't just the animals that were changing. The plants were also evolving, with the first seed-bearing plants emerging towards the end of the era. This innovation had a profound effect on the environment, as the new plants were able to colonize new areas and diversify the terrestrial ecosystem.

Of course, no discussion of the Paleozoic Era would be complete without mentioning the mass extinction events that punctuated this period. The most famous of these is the Permian–Triassic extinction event, which wiped out over 90% of all life on Earth. While the cause of this event is still debated, it is clear that it had a profound effect on the course of evolution, paving the way for the rise of the dinosaurs and the eventual emergence of mammals.

In conclusion, the Paleozoic Era is a period of time that is truly awe-inspiring in its scale and diversity. From the first multicellular organisms to the rise of the first land animals, this era is a testament to the incredible power of evolution and the ingenuity of life. Whether you are a student of science or simply an enthusiast of natural history, the Paleozoic Era is a time that is sure to captivate and inspire.

Mesozoic Era (252–66 Million Years Ago)

This incredible period of time spans over 180 million years and is home to some of the most iconic and beloved creatures in the history of life on Earth. From the mighty dinosaurs to the rise of the first mammals, the Mesozoic Era is a time of incredible change and evolution.

One of the most remarkable features of the Mesozoic Era is the dominance of the dinosaurs. These incredible creatures roamed the land for millions of years, evolving into a dizzying array of shapes and sizes. From the gentle giant Brachiosaurus to the ferocious T-Rex, the dinosaurs were a true testament to the power and diversity of evolution.

But it wasn't just the dinosaurs that were thriving during the Mesozoic Era. The seas were also teeming with life, with incredible creatures like ichthyosaurs and plesiosaurs ruling the waves. The skies, too, were alive with the sound of flying reptiles, including the iconic pterosaurs.

One of the most fascinating aspects of the Mesozoic Era is the way in which the continents were slowly drifting apart. This continental drift had a profound effect on the climate and environment, leading to the emergence of new habitats and the diversification of life. It also had a major impact on the formation of new mountain ranges, including the mighty Rocky Mountains.

But the Mesozoic Era was not just a time of incredible diversity and innovation. It was also a period of great turmoil, marked by a number of catastrophic events that had a profound effect on the course of evolution. The most famous of these events is the Cretaceous-Paleogene extinction event, which wiped out the dinosaurs and allowed for the rise of the mammals.

Despite the challenges that the Mesozoic Era faced, it was a time of incredible progress and innovation. From the evolution of the first flowers to the emergence of the first true mammals, this era set the stage for the incredible diversity of life that we see today.

In conclusion, the Mesozoic Era is a period of time that is truly awe-inspiring in its scale and diversity. From the reign of the dinosaurs to the rise of the mammals, this era is a testament to the incredible power of evolution and the ingenuity of life. Whether you are a student of science or simply an enthusiast of natural history, the Mesozoic Era is a time that is sure to captivate and inspire.

Cenozoic Era (66 Million Years Ago—Present)

This period, which spans from 66 million years ago to the present day, is an incredibly fascinating time in the history of our planet. From the extinction of the dinosaurs to the evolution of mammals, there are countless interesting and significant events that occurred during this era.

One of the most significant events of the Cenozoic Era was the extinction of the dinosaurs. This catastrophic event, which occurred around 66 million years ago, marked the end of the Mesozoic Era and paved the way for the rise of mammals. While the exact cause of the extinction is still debated, it is widely believed that a large asteroid impact was the main culprit.

With the extinction of the dinosaurs, mammals were able to take center stage. This led to the evolution of many new and diverse species, including primates, rodents, and ungulates. One of the most notable examples of this evolution was the rise of the hominids, which eventually led to the emergence of modern humans.

During the Cenozoic Era, the Earth's climate also underwent significant changes. At the beginning of the era, the climate was warm and tropical, but over time it gradually cooled. This led to the formation of polar ice caps and the development of more complex weather patterns.

In addition to these major events, there were also many smaller but no less interesting developments during the Cenozoic Era. For example, the era saw the evolution of many new plant species, including grasses, which would eventually become one of the most important food sources for many animals, including humans.

The Cenozoic Era also saw the rise of many new geological formations, including the Himalayan Mountains and the Great Barrier Reef. These formations not only provided new habitats for many species, but also helped to shape the Earth's climate and environment.

Overall, the Cenozoic Era is a fascinating period in the history of our planet. From the extinction of the dinosaurs to the evolution of mammals and the development of new geological formations, there are countless interesting and significant events that occurred during this era. So why not dive in and discover all that this incredible period has to offer?

Engaged Scientific and Intellectual Groups and Majors

Numerous scientific and intellectual disciplines, fields, categories, technologies, ideas, and concepts pertaining to environmentalism have emerged, encompassing:

Conservation Geography

This is the study of how human activities impact the environment and how to develop sustainable solutions to preserve biodiversity and natural resources.

Rural Geography

This is the field of study that focuses on understanding the functioning of villages and promoting sustainable technologies and practices to develop rural areas while preventing migration.

Urban Geography

This is the study of how cities and towns function and how to develop sustainable urban areas.

Climate Geography

This is the study of how climate changes over time and how humans impact it, with a focus on developing sustainable solutions to mitigate the effects of climate change.

Geomorphology

This is the study of the formation and evolution of landscapes, including landforms, soils, and water systems.

Biogeography

This is the study of the distribution of species and ecosystems over space and time, with a focus on understanding how human activities impact biodiversity.

Geospatial Technology

This is the application of Geographic Information Systems (GIS), remote sensing, and other geospatial technologies to analyze and solve environmental problems.

Environmental Geography

This is the study of how human activities impact the environment and how to develop sustainable solutions to protect natural resources and ecosystems.

Physical Geography

This is the study of the natural environment, including landforms, water systems, and climate, with a focus on developing sustainable solutions to preserve natural resources.

Human Geography

This is the study of human activity and how it affects the environment, with a focus on developing sustainable solutions to improve quality of life.

Geopolitics

This is the study of the political and economic interactions between countries and regions, with a focus on developing sustainable solutions to promote peace and cooperation.

Geoinformatics

This is the application of geospatial technology to analyze and visualize spatial data, with a focus on developing sustainable solutions to protect the environment.

Landscape Ecology

This is the study of the interactions between ecosystems and the landscape, with a focus on developing sustainable solutions to preserve biodiversity and natural resources.

Historical Geography

This is the study of how human activity has impacted the environment over time, with a focus on developing sustainable solutions to learn from past mistakes.

Social Geography

This is the study of how social factors, such as culture, politics, and economics, impact the environment, with a focus on developing sustainable solutions to promote social justice.

Environmental Health Geography

This is the study of how environmental factors impact human health, with a focus on developing sustainable solutions to improve public health.

Remote Sensing

This is the use of satellite and aerial imagery to study and monitor the environment, with a focus on developing sustainable solutions to protect natural resources and ecosystems.

Water Resources Geography

This is the study of water systems and how humans impact them, with a focus on developing sustainable solutions to protect water resources and ensure access to clean water.

Land Use Planning

This is the practice of designing and managing land use in a way that is sustainable and ecologically friendly, often with a focus on urban and regional development.

Ecology

This is a branch of biology that studies the relationships between living organisms and their environment.

Conservation Biology

This is a multidisciplinary field that aims to protect biodiversity and the natural environment.

Environmental Science

This field studies the natural world and the ways in which humans interact with and impact it.

Sustainability

This is the study of how to meet the needs of the present without compromising the ability of future generations to meet their own needs.

Climate Science

This field studies the Earth's climate system and the causes and effects of climate change.

Green Technology

This is the development and application of technology that is environmentally friendly.

Environmental Economics

This is the study of how economic activity impacts the environment and how to create economic policies that support sustainability.

Environmental Ethics

This is the branch of philosophy that deals with moral and ethical questions related to the environment and our relationship with it, and other issues including issues of biodiversity, animal rights, and ecological justice.

Environmental Policy

This refers to the laws, regulations, and guidelines that are put in place to manage the environment and address environmental issues.

Environmental Engineering

This is the application of engineering principles to design solutions to environmental problems, such as pollution control and waste management.

Environmental Health

This field studies how environmental factors can affect human health, including issues like air and water pollution, toxic chemicals, and climate change.

Environmental Education

This is the practice of teaching people about the environment and how to protect it, often with a focus on experiential learning and outdoor education.

Biophilia

This is the idea that humans have an innate connection to nature and that exposure to nature can have positive effects on mental health and well-being.

Deep Ecology

This is a philosophy that emphasizes the interconnectedness of all living things and advocates for a shift away from anthropocentric views of the environment.

Ecocentrism

This is a worldview that places the value of the environment above the interests of humans, and emphasizes the intrinsic value of nature.

Environmental Justice

This is the idea that environmental issues disproportionately affect marginalized communities, and that environmental policies should aim to address these inequalities.

Green Architecture

This is the design and construction of buildings that are environmentally sustainable and energy efficient.

Urban Ecology

This is the study of the relationships between organisms and their environment in urban areas, and how to design cities that are more sustainable and ecologically friendly.

Agroecology

This is the application of ecological principles to agriculture, with a focus on sustainability, biodiversity, and social justice.

Environmental Psychology

This is the study of how people interact with the environment and how the environment affects human behavior and well-being.

Ecological Economics

This is a field that seeks to integrate economic principles with ecological principles, with a focus on sustainability and the well-being of both humans and the environment.

Permaculture

This is a design system that aims to create sustainable human settlements by working with natural ecosystems and patterns.

Ecological Restoration

This is the practice of restoring damaged ecosystems to a more natural and functional state, often through the reintroduction of native species and the removal of invasive species.

Environmental Art

This is a form of art that is inspired by the natural world and often aims to raise awareness about environmental issues.

Systems Ecology

This is the study of the relationships between the different components of an ecosystem, and how changes in one component can affect the entire system.

Environmental Anthropology

This is the study of how human cultures interact with the natural environment, including issues of resource use, conservation, and sustainability.

Environmental History

This is the study of how human societies have interacted with the natural environment throughout history, and how these interactions have shaped both human culture and the environment itself.

Land Management

This is the practice of managing land resources in a sustainable and ecologically friendly way, often with a focus on conservation and restoration.

Marine Conservation

This is the practice of protecting and preserving marine ecosystems and the biodiversity they support, often through the creation of marine reserves and the implementation of sustainable fishing practices.

Geoengineering

This is the deliberate manipulation of the Earth's climate system to counteract the effects of climate change, such as through carbon capture and storage or solar radiation management.

Green Chemistry

This is the development and application of chemical processes and products that are environmentally sustainable and reduce or eliminate the use and generation of hazardous substances.

Environmental Criminology

This is the study of how environmental factors contribute to criminal behavior, and how environmental design and management can help prevent crime.

Environmental Law

This is the branch of law that deals with environmental issues, including pollution control, natural resource management, and the protection of endangered species and habitats.

Environmental Journalism

This is the practice of reporting on environmental issues, including climate change, pollution, and conservation.

Environmental Sociology

This is the study of how social factors and institutions interact with environmental issues, including issues of power, inequality, and social justice.

Environmental Planning

This is the practice of designing and managing land use in a way that is sustainable and ecologically friendly, often with a focus on urban and regional development.

Ecological Footprint Analysis

This is a tool for measuring the impact of human activity on the environment, including the use of resources, the generation of waste, and the emission of GHGs.

Environmental Data Science

This is the use of data analysis and modeling to understand environmental issues, including climate change, biodiversity loss, and pollution.

Environmental Philosophy

This is the study of the ethical and moral implications of human interactions with the natural world, and how to develop sustainable and just relationships with nature.

Environmental Science Communication

This is the practice of communicating scientific information about the environment to the public, often with a focus on promoting awareness and action.

Green Marketing

This is the marketing of products and services that are environmentally sustainable and socially responsible.

Environmental Finance

This is the management of financial resources to promote environmental sustainability and address environmental issues, such as through sustainable investing or green bonds.

Sustainable Tourism

This is the practice of tourism that is environmentally sustainable and socially responsible, often with a focus on ecotourism and cultural preservation.

Sustainable Agriculture

This is the practice of agriculture that is environmentally sustainable and socially responsible, often with a focus on organic farming, conservation agriculture, and agroforestry.

Environmental Monitoring

This is the practice of collecting and analyzing data on the environment, including air and water quality, biodiversity, and climate change.

Renewable Energy

Renewable energy is generated from sources that are naturally replenished, such as wind, solar, hydro, and geothermal energy. These sources are intended to serve as substitutes for fossil fuels.

Green Infrastructure

This is the design and implementation of infrastructure that is environmentally sustainable and socially responsible, often with a focus on urban and regional development.

Conservation Biology

This is the study of the preservation and management of biodiversity, including the protection of endangered species and the restoration of degraded ecosystems.

Ecological Agriculture

This is the practice of agriculture that is based on ecological principles, with a focus on sustainability, biodiversity, and the conservation of natural resources.

Environmental Governance

This is the system of rules, policies, and institutions that govern environmental issues, including issues of accountability, transparency, and public participation.

Environmental Modeling

This is the use of mathematical and computer models to simulate and predict environmental processes and phenomena, such as climate change, land use change, and biodiversity loss.

Green Entrepreneurship

This is the development of businesses and startups that are environmentally sustainable and socially responsible, often with a focus on innovation and technology.

Environmental Diplomacy

This is the practice of diplomacy that focuses on environmental issues, such as climate change, biodiversity, and natural resource management.

Green Urbanism

This is the design and management of cities that are environmentally sustainable and socially responsible, often with a focus on walkability, public transportation, and green spaces.

Environmental Archaeology

This is the study of human interactions with the environment in the past, including issues of resource use, landscape management, and climate change.

Do They Actually Work?

Nevertheless, the main question at hand is: how much do they actually work? Let us examine the statistics pertaining to them. The statistics about the Earth encompass a vast array of data related to various aspects of the planet such as its physical characteristics, climate patterns, ecosystems, human population, and more. Depending on the specific type of statistic being examined, the data may reveal information about the Earth's health, changes in its environment over time, the impact of human activities on the planet, and other insights that can help inform policies and actions aimed at promoting sustainability and protecting the planet. Some of the important aspects of the Earth's statistics include:

Natural Resources

The Earth's statistics on natural resources such as water, land, and minerals are essential for managing these resources sustainably and minimizing their depletion or degradation.

- Water: According to the United Nations, water scarcity affects more than 40% of the global population, and this number is expected to increase due to population growth and climate change. In addition, water pollution caused by industrial and agricultural activities poses a significant threat to human health and the environment.

- Land: Land degradation, desertification, and deforestation are major environmental problems that affect the health of the earth. The FAO estimates that about 33% of the world's land is moderately to severely degraded due to human activities such as overgrazing, land-use change, and unsustainable farming practices.
- Minerals: The extraction of minerals such as coal, oil, and gas has a significant impact on the environment, including air and water pollution, habitat destruction, and GHG emissions. In addition, the demand for rare earth minerals used in electronics and renewable energy technologies has led to environmental and social problems in some areas where they are mined.

Climate Patterns

Understanding the Earth's climate patterns is crucial for predicting and mitigating the impact of natural disasters such as hurricanes, floods, and droughts. Climate patterns are complex and vary across different regions of the Earth, but in general, statistics suggest that the Earth's climate is changing at an unprecedented rate. Here are some key statistics related to climate change:

- Global temperatures have risen by 0.99 [0.84 to 1.10] °C since the pre-industrial era, with 2020 being one of the three warmest years on record (IPCC 2021). This warming trend contributes to more frequent and severe heatwaves, impacting ecosystems, agriculture, and human health.
- Atmospheric concentrations of CO_2 have increased from approximately 280 parts per million (ppm) in the pre-industrial era to over 410 ppm today, primarily due to the burning of fossil fuels and land-use changes (IPCC 2021).
- The Arctic Sea ice has been declining rapidly since satellite records began in the late 1970s, with a loss of around 13.1% per decade (Stroeve et al. 2007). Also, NASA reports a significant reduction in Arctic Sea ice extent, with a decline of about 13.3% per decade. Glaciers worldwide are melting at an accelerated rate, contributing

to rising sea levels. GMSL has risen by approximately 20 cm since pre-industrial times, and are projected to continue to rise due to the melting of glaciers and ice sheets (IPCC 2021). The United Nations estimates a global mean sea level rise of about 3.3 mm per year, threatening coastal communities and low-lying islands.

Biodiversity

Biodiversity refers to the variety of life on Earth, including the diversity of species, ecosystems, and genetic diversity. WWF notes that the rate of species extinction is currently 1000 times higher than the natural background rate, primarily due to habitat destruction and climate change. Here are some statistics about the state of biodiversity on our planet:

- Species extinction rates: The current extinction rate is estimated to be 1000 to 10,000 times higher than the natural rate, with up to 1 million species at risk of extinction in the coming decades (Bongaarts 2019).
- Habitat loss: Habitat loss and degradation are the primary threats to biodiversity, with around 40% of the Earth's land surface converted for human use, and about 75% of the Earth's ice-free surface altered to some degree by human activity (Foley et al. 2005).
- Climate change: Climate change is also having a significant impact on biodiversity, with rising temperatures and changing precipitation patterns affecting ecosystems and species (Parmesan 2006).
- Ocean acidification: Ocean acidification, caused by the absorption of carbon dioxide into seawater, is affecting the growth and survival of marine organisms and ecosystems (Doney et al. 2009; Raven 2005).
- Genetic diversity: The loss of genetic diversity can have significant consequences for species and ecosystems, making them less adaptable to changing conditions (Allendorf et al. 2012).

Human Population

The health of the Earth is closely related to the human population and its impact on the planet. The Earth's population statistics can reveal important information about issues such as poverty, education, health, and resource consumption, which can inform policies and actions aimed at promoting sustainable development. Here are some statistics about the health of the Earth:

- According to the World Bank, the world's population has grown from 6.1 billion in 2000 to 7.9 billion in 2021, an increase of over 29%. Moreover, the current rate of population growth is about 1.05% per year. The growing human population puts pressure on natural resources and contributes to environmental issues such as deforestation, air and water pollution, and climate change. The demand for food (Smil 1994), water, and energy increases with population growth, leading to overconsumption and unsustainable use of resources. Rapid population growth can also lead to urbanization, which contributes to urban heat islands, air pollution, and other environmental problems.
- According to UNESCO, while the global literacy rate increased from 83% in 2000 to 86.3% in 2019; in 2000, about 115 million children and youth were out of school, compared to 258 million in 2020.
- Climate change, driven largely by human activities such as burning fossil fuels and deforestation, is causing global temperatures to rise and leading to more frequent and severe weather events, such as heat waves, droughts, floods, and storms.
- The concentration of CO_2 in the atmosphere has increased by about 50% since the pre-industrial era, primarily due to human activities, and is now higher than at any time in at least the last 800,000 years (USEPA 2016).
- The world's oceans have absorbed about 30% of the CO_2 emitted by human activities, leading to ocean acidification, which can harm marine life and ecosystems (Feely et al. 2009; Orr et al. 2005).
- The loss of biodiversity, driven by factors such as habitat destruction, overexploitation, and climate change, is accelerating.

- Air pollution, largely caused by burning fossil fuels and other human activities, is a major cause of premature death and disease worldwide. In 2019, an estimated 4.2 million premature deaths globally were attributed to outdoor air pollution (Burnett et al. 2018; Lelieveld et al. 2019).
- Water scarcity affects at least 1.8 billion people globally, and this number is expected to increase as a result of population growth, climate change, and unsustainable water use (Canton 2022).

Environmental Impact

The statistics regarding the environmental impact of the Earth are concerning. Tracking statistics related to human activities such as GHG emissions, waste generation, and energy consumption can help assess the impact of these activities on the environment and inform policies aimed at reducing them. Here are some key points related to GHG emissions, waste generation, and energy consumption:

- GHG Emissions: According to IPCC, the amount of GHGs in the atmosphere that contribute to global warming has increased by 0.59 watts per square meter between 1750 and 2019. This is higher than what was estimated in a previous report covering the years 1750 to 2011 (IPCC 2021) (Fig. 1.1).
- Waste Generation: The World Bank estimates that global waste generation will increase by 70% by 2050, reaching 3.4 billion tonnes per year. Most of this waste will end up in landfills, where it can contribute to GHG emissions, as well as other environmental problems such as groundwater contamination and soil pollution (Kaza et al. 2018).
- Energy Consumption: According to IEA, global energy consumption is projected to increase by 25% by 2040, with fossil fuels continuing to be the dominant energy source. This is despite the growing availability and affordability of renewable energy sources like solar and wind power (IEA 2019).

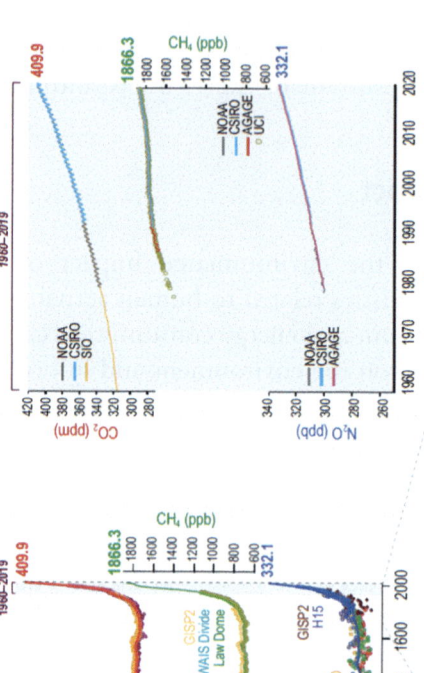

Fig. 1.1 GHG emissions (IPCC 2021)

Ocean Health

The statistics paint a concerning picture of the health of the Earth's oceans. They can provide insights into issues such as overfishing, ocean acidification, and plastic pollution, which can have significant impacts on marine ecosystems and human populations that rely on them. Here are a few key statistics related to overfishing, ocean acidification, and plastic pollution:

- Overfishing: According to FAO, around 34% of fish stocks are currently being overfished, while 60% are being fished at their maximum sustainable level. In some regions, overfishing has led to the collapse of entire fish populations, such as the cod stocks off the coast of Newfoundland in Canada.
- Ocean Acidification: The oceans are absorbing excess carbon dioxide, leading to ocean acidification. The pH of the ocean is currently around 8.1, which is about 0.1 lower than it was before the Industrial Revolution (Raven 2005). This reduction in pH, although it seems small, represents a significant increase in ocean acidity due to increased atmospheric CO_2 levels from human activities since the Industrial Revolution. This increase in acidity is primarily due to the absorption of carbon dioxide from the atmosphere, which forms carbonic acid in the ocean.
- Plastic Pollution: Around 8 million metric tons of plastic enter the ocean every year, according to UNEP (2016). Plastic pollution is particularly harmful to marine wildlife, with an estimated 1 million seabirds and 100,000 marine mammals dying each year as a result of plastic pollution (Wilcox et al. 2015).

Land Use

The statistics regarding land use and its impact on the health of the Earth are complex and varied. They such as agricultural practices and urbanization, can provide insights into the impact of human activities on natural habitats and the health of ecosystems. However, there are some general trends that can be observed:

- Deforestation: Forests, crucial for carbon sequestration, are being rapidly depleted. FAO estimates that 10 million hectares of forest are lost annually.
- Agricultural practices: Agriculture is the largest user of land globally, and the way we use land for agriculture has significant impacts on the health of the Earth. For example, industrial agriculture often relies heavily on pesticides and fertilizers, which can harm soil quality, water quality, and biodiversity. It can also lead to deforestation, which contributes to climate change. On the other hand, sustainable agriculture practices, such as organic farming, agroforestry, and regenerative agriculture, can promote soil health, biodiversity, and carbon sequestration.
- Urbanization: As the world becomes more urbanized, there is a growing demand for land to support cities and their infrastructure. This can lead to the conversion of natural habitats, such as forests and wetlands, into urban areas. This can have negative impacts on biodiversity, soil quality, and carbon sequestration. However, urbanization can also promote more efficient land use and reduce the footprint of human activities on the Earth's ecosystems.

Natural Disasters

Statistics on natural disasters such as earthquakes, wildfires, and land subsidence can inform policies and actions aimed at mitigating their impact and improving preparedness and response. The statistics related to earthquakes, wildfires, and land subsidence provide insights into the health of the Earth. Here are some key points:

- Earthquakes: Earthquakes are a natural part of the Earth's tectonic processes and can release built-up pressure in the Earth's crust, although some human activities such as drilling and hydraulic fracturing can also induce earthquakes by altering the stress and pressure

in the subsurface. Earthquakes can cause extensive damage to infrastructure and buildings, leading to economic losses and loss of life.
- Heatwaves: According to WMO, the number of heatwaves has increased in frequency, length, and intensity since the 1950s, and this trend has continued into the twenty-first century.
- Wildfires: Wildfires can cause significant damage to ecosystems and contribute to climate change by releasing carbon dioxide into the atmosphere. However, wildfires are also a natural part of many ecosystems, and some plant species have evolved to depend on periodic fires to regenerate.
- Droughts: Droughts have affected more people than any other natural disaster over the past century, according to the WMO. In recent years, severe droughts have occurred in many parts of the world, including the western United States, Australia, and South Africa. In 2021, Brazil experienced its worst drought in nearly a century, causing widespread impacts on agriculture, hydropower generation, and biodiversity.
- Hurricanes: the number of hurricanes globally has remained relatively stable over the past several decades, but there has been an increase in the frequency of the most powerful hurricanes (category 4 and 5) in some regions, such as the North Atlantic. The 2020 Atlantic hurricane season was one of the most active on record, with 30 named storms and 13 hurricanes, and it caused significant damage and loss of life in several countries, including the United States, Nicaragua, and Honduras.
- Warm/hot extremes: according to IPCC, the frequency and intensity of warm/hot extremes have increased globally over the past century, with a greater than 90% probability that human-caused emissions have contributed to this trend. In 2021, several countries experienced record-breaking temperatures, including Canada, which recorded a temperature of 49.6 °C (121.3 °F) in British Columbia, and Siberia, which reached 38 °C (100 °F) in the Arctic town of Verkhoyansk.
- Cold extremes: according to recent studies, there is evidence of a decrease in the frequency and intensity of cold extremes, which refers to a decrease in colder and/or fewer cold days and nights and cold spells/cold waves over most land areas. However, it's important to note that this does not necessarily mean that cold weather events are

no longer occurring, but rather that they are becoming less frequent overall.
- Heavy precipitation: The frequency and intensity of heavy precipitation events have increased globally over the past century, with a greater than 90% probability that human-caused emissions have contributed to this trend, according to the IPCC. In 2021, several countries experienced devastating floods, including Germany and Belgium, which saw record-breaking rainfall that caused widespread damage and loss of life.
- Extreme sea levels: as mentioned before, GMSL has risen by about 20 cm (8 in.) since 1900, with a significant acceleration in the rate of rise in recent decades, according to the IPCC. In 2021, Hurricane Ida caused a storm surge that inundated parts of Louisiana and Mississippi with extreme sea levels, causing significant damage and loss of life.
- Land subsidence: Excessive withdrawal of groundwater is a problem that occurs in some parts of the world. This issue is particularly prevalent in regions with arid and semi-arid climates, where the demand for water exceeds the natural recharge rate of aquifers. For example, in parts of India, China, and the United States, groundwater is being extracted at rates that exceed the rate of natural replenishment, leading to depletion of the resource. Additionally, in some areas, excessive use of groundwater for irrigation has led to land subsidence, which can result in infrastructure damage and increased flood risk. While this is not a universal issue, it is a growing concern in many regions and requires effective management and conservation strategies to ensure the long-term sustainability of groundwater resources.

Air Quality

Tracking statistics related to air quality can help assess the impact of human activities such as transportation and industrial production on the environment and human health. Studies have shown that air pollution is linked to a range of health problems, including respiratory diseases, cardiovascular diseases, and cancer. According to WHO, air pollution is responsible for an estimated 7 million premature deaths annually.

The uncontrolled emissions from vehicles, factories, and other sources contribute to air pollution, which has numerous harmful effects on the environment and human health.

- Transportation: Transportation is a significant contributor to air pollution, with the burning of fossil fuels in cars, trucks, and airplanes releasing pollutants such as nitrogen oxides, particulate matter, and carbon monoxide into the air.
- Industrial production is also a major source of air pollution, with emissions from factories, power plants, and other facilities contributing to both local and global air pollution. The types and amounts of pollutants vary depending on the industry and the specific processes used, but common air pollutants from industrial production include CO_2, SO_x, PM_{10}, SO_2, NO_x, VOCs, and heavy metals.

Conclusion

Overall, the statistics paint a concerning picture of the state of the Earth and the negative impacts of human activities on its health. The realization that human activities are adversely affecting the bio-physical environment, once a concern primarily of industrial nations, has now become a global awareness (Dietz and Rosa 1994). Either the various scientific and intellectual disciplines, fields, technologies, ideas, and concepts mentioned earlier seem ineffective, or the degree of Earth's destruction is substantial.

Unsustainable land use practices, significant changes to biodiversity, pressure on natural resources, and the environmental impact of human activity all require urgent action. The health of the Earth's oceans is also in jeopardy, and natural disasters, while a natural part of the Earth's processes, require effective management and mitigation. In addition, human-induced climate change is a major issue that must be addressed to ensure a healthier future for the planet and its inhabitants. This challenge involves much more than simply studying changes in the global climate system; it encompasses the Earth System in its entirety, including physical, bio-geochemical, and societal processes (Ehlers and Krafft 2006).

Therefore, it is true that significant efforts have been made in environmental protection and conservation, but the challenges facing the Earth remain crucial. In upcoming seasons, I will discuss why we have faced these kinds of problems and provide some of the scientific principles according to the latest scientific findings that underlie the Earthmate philosophy emphasizing the need for a holistic and collaborative approach to environmental management.

References

Allendorf, F. W., Luikart, G. H., & Aitken, S. N. . (2012). *Conservation and the genetics of populations.* John Wiley & Sons.

Bongaarts, J. (2019). *Summary for policymakers of the global assessment report on biodiversity and ecosystem services of the Intergovernmental Science-Policy Platform on Biodiversity and Ecosystem Services.* Wiley Online Library.

Burnett, R., Chen, H., Szyszkowicz, M., Fann, N., Hubbell, B., Pope III, C. A., ... Weichenthal, S. (2018). Global estimates of mortality associated with long-term exposure to outdoor fine particulate matter. *Proceedings of the National Academy of Sciences,* 115(38), 9592–9597.

Canton, H. (2022). *The Europa directory of international organizations.* Routledge.

Dietz, T., & Rosa, E. A. (1994). Rethinking the environmental impacts of population, affluence and technology. *Human ecology review,* 1(2), 277-300.

Doney, S. C., Fabry, V. J., Feely, R. A., & Kleypas, J. A. (2009). Ocean acidification: the other CO_2 problem. *Annual review of marine science,* 1, 169-192.

Ehlers, E., & Krafft, T. (2006). *Earth system science in the anthropocene.* Springer.

Feely, R. A., Doney, S. C., & Cooley, S. R. . (2009). Ocean acidification: Present conditions and future changes in a high-CO_2 world. *Oceanography,* 22(4), 36-47.

Foley, J. A., DeFries, R., Asner, G. P., Barford, C., Bonan, G., Carpenter, S. R., . . . Gibbs, H. K. (2005). Global consequences of land use. *science,* 309(5734), 570–574.

IEA. (2019). *World Energy Outlook 2019*. Paris: IEA.
IPCC. (2021). *Climate Change 2021: The Physical Science Basis. Contribution of Working Group I to the Sixth Assessment Report of the Intergovernmental Panel.* Cambridge, United Kingdom and New York, NY, USA: Cambridge University Press.
Kaza, S., Yao, L., Bhada-Tata, P., & Van Woerden, F. (2018). *What a waste 2.0: a global snapshot of solid waste management to 2050*. World Bank Publications.
Lelieveld, J., Klingmüller, K., Pozzer, A., Pöschl, U., Fnais, M., Daiber, A., & Münzel, T. (2019). Cardiovascular disease burden from ambient air pollution in Europe reassessed using novel hazard ratio functions. *European heart journal*, 40(20), 1590-1596.
Orr, J. C., Fabry, V. J., Aumont, O., Bopp, L., Doney, S. C., Feely, R. A., ... Joos, F. (2005). Anthropogenic ocean acidification over the twenty-first century and its impact on calcifying organisms. *Nature*, 437(7059), 681–686.
Parmesan, C. (2006). Ecological and evolutionary responses to recent climate change. *Annual Review of Ecology, Evolution, and Systematics*, 37, 637-669.
Raven, J. (2005). *Ocean acidification due to increasing atmospheric carbon dioxide*. London, UK: The Royal Society.
Smil, V. (1994). How many people can the earth feed? *Population and Development Review*, 255–292.
Stroeve, J., Holland, M. M., Meier, W., Scambos, T., & Serreze, M. (2007). Arctic sea ice decline: Faster than forecast. *Geophysical research letters*, 34(9).
UNEP. (2016). *Marine plastic debris and microplastics – Global lessons and research to inspire action and guide policy change*. Nairobi: United Nations Environment Programme.
USEPA. (2016). *Climate Change Indicators: Atmospheric Concentrations of Greenhouse Gases*. Washington DC, USA: United States Environmental Protection Agency.
Wilcox, C., Van Sebille, E., & Hardesty, B. D. (2015). Threat of plastic pollution to seabirds is global, pervasive, and increasing. *Proceedings of the National Academy of Sciences*, 112(38), 11899-11904.

2

Earthmate Philosophy

Introduction

Numerous ancient and modern theories, philosophies, beliefs, and considerable hypotheses from various cultures, whether religious or nonreligious, have sought to elucidate and emphasize a fundamental unity and interconnectedness within the universe. Here are various philosophical concepts and ideas, including Spinoza's "Deus sive Natura", Pantheism, Panentheism, and Naturalistic pantheism, as well as Neoplatonism's concepts of The One and Emanation, the Stoic notion of Logos, Schelling's Naturphilosophie, Monism, Holism, Leibnizian Metaphysics' Monad, the Absolute in German Idealism, particularly in Hegelian Dialectics, the concept of élan vital in Vitalism and Organicism against Mechanism, the concept of Universal Mind or Universal Consciousness in New Age Spirituality, and the Anima Mundi (World Soul), among others. These concepts may find their origins in various religious beliefs, such as Waḥdat al-Wujūd (unity of existence) in Islamic mysticism, Brahman-Atman in Advaita Vedanta, Tao in Taoism, Śūnyatā (emptiness) in Mahayana Buddhism, Great Spirit in Native American Spirituality, and the concept of immanence, among others. Regardless

© The Author(s), under exclusive license to Springer Nature Switzerland AG 2025
A. Faridi Masouleh, *Earthmate*, https://doi.org/10.1007/978-3-031-82556-9_2

of their unique functions, we observe common threads in them underscore a concept known as "Unity Despite Plurality," which serves as the foundational philosophy of Earthmate.

Moreover, within empirical sciences, particularly in Earth-centric disciplines here, various concepts, approaches, issues, and technologies have surfaced, emphasizing the significance of Interaction and interconnectedness of the "Unity Despite Plurality" philosophy elucidated by the aforementioned concepts. Among the numerous concepts, a selection includes ESS, Ecosystem, Gaia hypothesis, Sustainable Development, Global temperature record, Climate change, Global warming, Synoptic scale meteorology, Sea level rise, GHG emissions, Ocean Acidification, Deforestation and Reforestation, carbon pool or Carbon sink, Planetary Boundaries, Complex system and the concept of Emergence in this context, Resilience of Socio-Ecological Systems, Climate Justice and Equity, Resilience and Vulnerability, Adaptation vs. Mitigation, Carbon Footprint, Climate Change Feedback Loops, and Tipping Points. These and a multitude of other concepts will be explored in upcoming seasons.

Deus sive Natura

Spinoza's ancestors were Sephardic Jews who, in the sixteenth century, established themselves along the borders of Spain and Portugal. They did so to engage in a prosperous trade within a relatively secure region of the Iberian Peninsula, protected by Muslim rulers (Scruton 2002). When Baruch Spinoza (1632–1677) was just six years old, his mother, Ana Débora, Miguel's second wife, passed away (Nadler 2001) and Spinoza's father, Miguel, passed away in 1654, when Spinoza was 21 years old. Spinoza received a conventional Jewish upbringing, enrolling at the Keter Torah yeshiva of the Amsterdam Talmud Torah congregation under the guidance of the erudite and orthodox senior Rabbi Saul Levi Morteira (Nadler 1999). He also studied under the somewhat less orthodox Rabbi Manasseh ben Israel (Scruton 2002). Influenced by thinkers such as Moses ben Maimon (Maimonides, 1135–1204), who drew from Aristotle (384–322 BC), Al-Farabi (c. 870–950), Avicenna or Ibn Sina

(980–1037), as well as his contemporary Averroes or Ibn Rushd (1126–1198), and Ibn Arabi (1165–1240)—renowned for explicitly formulating the concept of "Wahdat ul-Wujud" ("Unity of Being") (Landau 2013)—played a significant role in the resurgence of Aristotelian philosophy. Their efforts greatly influenced the course of medieval theology towards an Aristotelian trajectory (Scruton 2002; Yalom 2012). Despite being frequently labeled an "atheist" by his contemporaries, Spinoza's works never explicitly argue against the existence of God (Stewart 2007; Simkins 2014). Nonetheless, in June 1678, just over a year after Spinoza's death, the States of Holland banned his entire works, since they "contain very many profane, blasphemous and atheistic propositions." The prohibition included the owning, reading, distribution, copying, and restating of Spinoza's books, and even the reworking of his fundamental ideas (Israel 1996). Not long thereafter, his works found their place on the Catholic Church's Index of Forbidden Books (Totaro 2015).

Spinoza's philosophy is primarily expounded in two key works: the Theologico-Political Treatise and the Ethics. His remaining writings are incomplete expressions of ideas that found their full form in these aforementioned books. Others are not directly related to Spinoza's own philosophy, like The Principles of Cartesian Philosophy and The Hebrew Grammar. Additionally, a collection of letters he left behind sheds light on his ideas and potential motivations. While the Theologico-Political Treatise was published in Spinoza's lifetime, the magnum opus of his philosophical system, the Ethics, was released posthumously in the year of his passing. This monumental work challenged Descartes's philosophy of mind–body dualism, securing Spinoza's legacy as one of the most influential thinkers in Western philosophy.

One of the important philosophical concepts developed by this 17th-century Dutch philosopher was "Deus sive Natura" which may be rooted in "Wahdat ul-Wujud". It is Latin for "God or Nature." According to Spinoza, God and Nature are synonymous, suggesting that the universe and all its constituents are expressions of a single, unified substance. This substance, which Spinoza called "God" or "Nature," encompasses everything in existence. This concept may give nature as a whole a divine meaning in essence and prevents us from hurting it.

Absolute Idealism

German idealism arose in Germany, stemmed from the ideas of Kant (1724–1804) (Beiser 2002). Within the realm of German idealism, one classification separates thinkers into transcendental idealists, linked with Kant and Fichte, and absolute idealists, linked with Schelling and Hegel (Dunham et al. 2011). This distinction encompasses the best-known figures in this philosophical movement. German idealism was intimately connected with both Romanticism and the revolutionary ideals of the Enlightenment.

Kant's work aimed to reconcile two prominent 18th-century philosophical schools: rationalism and empiricism. Empiricism, championed by Hume (1711–1776), introduced skepticism, which Kant aimed to counter (Dudley 2014). He proposed that while we rely on empirical objects for knowledge, we can a priori investigate the framework of our thoughts, setting the limits of possible experience. He termed it "critical philosophy", more focused on critiquing theoretical boundaries than establishing positive doctrine. Immanuel Kant posited that the human mind cannot directly apprehend the external world in its true essence; rather, our perception is filtered through inherent a priori categories and concepts. Kant terms these as "transcendental" since they are indispensable for any experience, shaping and arranging our understanding of the world. His "transcendental idealism" means the mind significantly shapes our experience, perceiving phenomena in time and space through mental categories. Kant's transcendental idealism comprises two key elements. Firstly, it posits that the human mind isn't merely a passive receiver of sensory input but actively molds our perception of the world. Secondly, it asserts that the true nature of reality remains ultimately beyond our reach, as our experience is filtered through the frameworks of our own minds. Kant confined knowledge to objects within the realm of potential experience. Nonetheless, his three most notable successors would challenge these strict boundaries (Guilherme 2010).

Johann Gottlieb Fichte (1762–1814) expanded on Kant's ideas, emphasizing the interplay between the self and the external world. Fichte's philosophy, akin to Kant's, eliminates the notion of a thing-in-itself. He posited that our perceptions stem from the "transcendental ego", or the knowing subject, negating the existence of an external thing-in-itself. According to Fichte, the subject generates the external object or non-ego. He asserted that this truth is evident through intellectual intuition, where reason allows for immediate apprehension. One of the last important point is thar it was actually Fichte, not Hegel, who first introduced the concept of thesis-antithesis-synthesis, a notion frequently misattributed to Hegel (Solomon 1985).

Schelling (1775–1854) argued that Fichte's "I" necessitates the Not-I. This means that subjective representations are essentially the same as the extended objects external to the mind. The concept of "absolute identity" or "indifferentism", there is no distinction between the subjective and the objective, or between the ideal (concepts or mental representations) and the real (external objects or reality itself). This means that, according to Schelling, the subjective and objective aspects of reality are ultimately inseparable, leading to a perspective where the distinctions between our mental constructs and the external world become less inflexible. There is an important philosophical approach named Naturphilosophie or philosophy of nature developed by Schelling which more details will be tabled in the next section.

Georg Wilhelm Friedrich Hegel (1770–1831) built upon Kant's philosophy, contending that the insoluble contradictions outlined in Kant's Critique of Pure Reason extended more broadly to reality itself. According to Hegel, true understanding of being can only be achieved through a comprehensive whole (das Absolute) known as the Absolute. He argued that for the thinking subject (human reason or consciousness) to truly know its object (the world), there must be, in some sense, a unity between thought and being. Without this unity, the subject would be unable to access the object, leading to uncertainty about our knowledge of the world. Recognizing the limits of abstract thought, Hegel explored how historical developments shaped various philosophies and modes of cognition. In "The Phenomenology of Spirit," he traced the evolution of self-consciousness through history and underscored the role of others in its awakening, introducing the pivotal concepts of historical importance and intersubjectivity to metaphysics and philosophy. Additionally, Hegel proposed his concept of absolute spirit, which he saw as a transcendent evolution of the traditional notion of God. He revered Spinoza for transforming the anthropomorphic concept of God into that of an underlying substance, a concept aligned with Hegel's idea of absolute knowing, asserting, "You are either a Spinozist or not a philosopher at all." (Hegel 1892) In German Idealism, especially in the philosophy of G. W. F. Hegel, the "Absolute" represents the ultimate, the highest, most comprehensive, all-encompassing reality or truth. It's the highest, most comprehensive concept that encompasses and integrates all concepts, forms, and phenomena representing the ultimate unity of everything that exists. Contrary to common belief, the concept of the Absolute is not unique to Hegel; its initial appearance can be traced back to the writings of Nicholas of Cusa. Hegel's interpretation of the Absolute was influenced and further refined in response to the ideas of his contemporary, Friedrich Wilhelm Joseph Schelling (Inwood 1992). Absolute idealism, associated primarily with Schelling and Hegel, also encompasses thinkers like Josiah Royce and the British idealists. According to Hegel, the Absolute is not a static entity, but a dynamic process of becoming, involving a dialectical movement of thesis-antithesis-synthesis. Hegel's dialectical process involves the movement from a thesis to its negation (antithesis) and then the synthesis of these opposing elements. This process leads to a higher level of truth and understanding. The absolute idealist stance sets itself apart from Berkeley's subjective idealism, Kant's transcendental idealism, and the post-Kantian transcendental idealism advocated by Fichte and early Schelling (Limnatis 2008).

Naturphilosophie

Naturphilosophie, or "philosophy of nature," was a philosophical approach developed by German philosopher Friedrich Wilhelm Joseph Schelling in the late 18th and early nineteenth centuries. It aimed to grasp the entirety of nature and delineate its overarching theoretical framework. It outlines a complex philosophical perspective on the nature of reality, focusing on the interplay between organic and inorganic elements in the natural world. It argues that nature is fundamentally dynamic and productive, with an infinite creative potential that requires inhibition to result in finite products. The story started when Fichte sought to establish that the entire structure of reality stems from self-consciousness. Schelling, building on Fichte's ideas, argued that nature possesses independent reality. Historically, Schelling observes, there have been two prevailing perspectives: one where the object is seen as absolute, encompassing the subject (which he labels as Spinozism), and another where the subject is regarded as absolute, embodying both content and form, essentially a "subject-in-itself" (akin to Fichte's idealism) (von Schelling 2004). However, he found Fichte's doctrines incomplete, as they closely tied the ultimate ground of the universe to finite, individual Spirit. This perspective risked undermining the reality of the natural world by leaning towards subjective idealism. According to this view, Fichte failed to integrate his system with Kant's aesthetical and teleological view of nature. Naturphilosophie offers one possible explanation for the unity of nature, considering both objective reality and self-consciousness as equally genuine. It suggests that philosophy encompasses both the study of nature and transcendental idealism, forming a complementary whole. Schelling emphasized the interconnectedness of all natural phenomena and sought to bridge the gap between nature and spirit.

Collectivism Versus Individualism

Collectivism and individualism are two contrasting philosophical and cultural perspectives that address the balance between the interests of the individual and the interests of the group or society as a whole. G. W. F. Hegel was a notable figure in the realm of Collectivism, while John Locke gained prominence in the sphere of Individualism. While the initial collectivization arose from historical contingencies like the village leader's preference and specific social-economic conditions, it has endured due to the formation of a common interest group among villagers (Hou 2013). The pre- and non-Marxian socialists (e.g. Owen, St. Simon, Fourier, etc.) have inherently embraced a collectivist concept of humanity (László 2013); and Locke's individualism extends beyond asserting natural freedom and equality, contending that individuals inherently own their person and abilities, being beholden to society for nothing; however, this form of individualism ultimately leads to a form of collectivism, as it necessitates the accumulation of property for full realization, achievable only by some and at the expense of others' individuality, emphasizing the supremacy of civil society over each individual (Macpherson 1962).

While collectivism versus individualism is one way to conceptualize cultural differences between the West and the East, it is important to recognize that these differences are not absolute and that there is a great deal of diversity within each region. It is important to be cautious when making cultural generalizations, as they can be oversimplified and may not accurately reflect the complexity of a culture. Understanding individualism and collectivism as the inclination to prioritize oneself as an individual or as part of a group can be valuable. However, using broad measures of this concept that don't consider specific cultural norms and values may lead to misinterpretations and might not be very effective in predicting cultural variations in decision-making and behavior (Briley and Wyer 2001). For example, while many Asian countries such as China and Japan are collectivistic cultures, there are also individualistic cultures in Asia, such as Hong Kong and Singapore. Similarly, while the United States is often cited as the most individualistic country, there are also collectivistic cultures in the West, such as Native American cultures. Midcentury historians depicted individualism as a pragmatic stance, advocating for petit-bourgeois values and an unhindered pursuit of material success; they contended that Americans, whether for better or worse, diverged from the idealism and audacious collective endeavors observed in European political movements. Conversely, Alex Zakaras underscores the utopian essence of American individualism, consistently aligning it with ideals of agrarian virtue, natural markets, and a classless meritocracy, weaving together personal freedom with concepts of divine justice and human redemption (Zakaras 2022).

Let's focus on their prominent attributes. Considering Focus, collectivism emphasizes the importance of the collective or group. It prioritizes the welfare and goals of the community, society, or nation over those of individuals. In collectivist societies, individuals are often expected to conform to societal norms and expectations. While individualism puts the individual at the center. It stresses personal freedom, autonomy, and self-expression. Individualists believe that each person should be free to pursue their own goals and make their own choices, even if those choices go against societal norms. Considering Interdependence, collectivism values interdependence and cooperation among members of a group. It often promotes a sense of solidarity and shared responsibility. In collectivist cultures, people may rely on extended family or community support networks. While individualism encourages self-reliance and independence. It emphasizes the individual's ability to make decisions for themselves and take care of their own needs. Considering Identity and Roles, collectivism places a strong emphasis on group identity. People in collectivist societies may define themselves primarily in terms of their family, community, or nationality. Social roles and expectations are often clearly defined. While individualism focuses on personal identity. Individuals in individualist cultures may define themselves more in terms of their unique attributes, achievements, and aspirations. Social roles may be more flexible and subject to personal choice. Considering Wealth Distribution, collectivism tends to support more centralized control of resources with the aim of reducing economic disparities. This may manifest in various forms, such as socialist or communist ideologies. While individualism advocates for a free-market system where individuals have the freedom to pursue their economic interests with minimal government intervention. It generally accepts a greater degree of economic inequality as a byproduct of individual freedoms. Considering Political Systems, collectivism is often associated with ideologies like socialism, communism, and some forms of tribal or communal governance. These systems often involve significant state control or ownership of resources. While individualism Correlates with capitalist systems, where private ownership, competition, and free markets play a central role. It advocates for limited government intervention in the economy. Considering social systems, in general methodological individualism, which is non-atomistic, adopts a bottom-up and emergentist approach, whereas sociological holism, ontological collectivism and heteronomy, operates on the basis of a top-down logic (Magnani et al. 2015). Considering Cultural Examples, collectivism examples include cultures like China, Japan, Korean culture, and many Indigenous communities where the emphasis is on group harmony and conformity. While individualism is

Commonly found in Western cultures like the United States, Canada, and much of Western Europe, where individual rights and personal freedoms are highly valued (Im 2019).

Societies incorporate elements of both collectivism and individualism to varying degrees. It's important to note that no society is purely collectivist or individualist; they exist on a spectrum, and cultural, historical, and economic factors influence where a society falls on that spectrum. Of course, the notion that individualism and collectivism represent opposite ends of a scale for arranging states and theories of the state, irrespective of their stage of social development, is superficial and misleading; Locke's individualism demands the supremacy of the state over the individual, indicating that it's not a matter of more individualism leading to less collectivism, but rather, the more thorough-going the individualism, the more comprehensive the collectivism, exemplified supremely by Hobbes's theory (Macpherson 1962). Additionally, individuals within a society may have their own unique perspectives that don't neatly align with the predominant cultural values. The transition from the Holocene to the Anthropocene, marked by the Industrial Revolution around 1800, brought about significant geological and cultural shifts, with a prevailing individualistic orientation and exploitation of the environment in developed nations, contrasting with the enduring collectivist mindset of Indigenous cultures, highlighting the potential for coexistence and mutual appreciation of diverse knowledge systems in this epoch (McIntyre-Mills et al. 2017). Now, the vital issue is striking a balance between the needs or rights of individuals with their individual duties or responsibilities vis-à-vis the local and international community and the earth, which is a common goal. This balance is crucial because individual well-being, the well-being of the community, and sustainable environment are intertwined and interdependent. It is a vital issue in many aspects of society especially when it generalizes to a greater and deeper issue or terrestrial interests as the big society of human beings. Terrestrial interests, referring to the concerns and priorities related to life on Earth, encompass a wide range of factors. These can include economic considerations, such as resource allocation and distribution, environmental concerns like conservation and sustainable practices, social justice issues involving access to basic needs and opportunities, and political considerations like governance and representation. Striking this balance among individuals, the community, and terrestrial interests is not always straightforward. It requires thoughtful decision-making, ethical considerations, and a willingness to adapt in the face of changing circumstances. It also involves active participation from individuals, communities, and institutions at various levels. Here, we can introduce the concept of the 'common good,' recognizing the significant influence of the market in shaping human society and its pivotal role in this context. Smith doubted the efficacy of those who traded for the public good, and Hayek argued that justice pertains only to human conduct, not market outcomes, concluding that the 'common good' is an abstract order that doesn't dictate specific outcomes but aims to maximize overall chances in the long term (Infantino 2014). Therefore, the ultimate aim is to forge a society where individuals, communities, and institutions are enabled to pursue their rational aspirations and exercise their rights considering the market, all the while recognizing, considering, and respecting the interconnected destiny of humanity and the earth. It can unite us all but entails cultivating a profound responsibility towards the earth and promoting sustainable personal and societal growth, with the understanding that our present prosperity paves the way for future generations to thrive equitably.

Organicism Versus Mechanism

Organicism and Mechanism are contrasting philosophical perspectives used to understand the nature of complex systems, including living organisms.

Let's focus on their vital attributes. Considering the Core Idea, Organicism views systems as integrated wholes, where the properties and behaviors of the whole cannot be reduced to or explained solely by the properties and behaviors of their individual parts. Mechanism, on the other hand, views systems as akin to machines, where the behavior of the whole can be entirely explained by understanding the behavior of its individual components. Considering the metaphorical interpretation of them, Organicism is often likened to a living organism, where every part plays a role in the functioning of the whole. Mechanism, on the other hand, is often compared to a clockwork mechanism, where precise interactions of parts determine the overall behavior. Considering the Approach, Organicism emphasizes the interdependence and interconnectivity of elements within a system and sees systems as dynamic and constantly evolving. While Mechanistic thinking seeks to break down complex systems into their constituent parts and analyze how they interact based on predefined rules or laws. Considering Historical Influence, Organicism has deep historical roots, with philosophers like Aristotle contributing to its development. It was also influential in the Romantic era and had a significant impact on early ecological thinking. Mechanistic thinking gained prominence during the Scientific Revolution, with figures like Descartes and Newton advocating for this reductionist approach to understanding the natural world. Considering Reductionism, Mechanism is often associated with reductionism, the idea that complex phenomena can be understood by breaking them down into simpler, more fundamental components. Organicism, on the other hand, is skeptical of complete reductionism, emphasizing the importance of understanding the whole. Considering Interconnectedness, Mechanism focuses on the cause-and-effect relationships between individual components, while Organicism highlights the interconnectedness and interdependence of elements within a system. Considering Application, Mechanism is commonly used in physics and engineering, where precise control and prediction of behavior are essential. Organicism is often applied to biological and ecological systems, where understanding the holistic nature of organisms and ecosystems is crucial.

By looking at some examples, we can see that biological systems, ecosystems, and even social systems are often approached from an organicist perspective. While mechanistic approach is commonly used in physics and engineering, where understanding the interactions of individual components is crucial to predict the behavior of the whole. In practice, many systems in the natural world exhibit elements of both organicism and mechanism, and the choice of perspective depends on the specific questions being asked and the level of analysis being applied.

Unity Despite Plurality (UDP)

UDP is a foundational philosophy of Earthmate that finds its roots in various ancient and modern philosophies, both religious and non-religious, from different cultures around the world. Unlike philosophers like Agius (1989), who base their philosophical theories like a relational ethical theory of intergenerational ethics (which is the emphasis of Earthmate Philosophy) on metaphysics, Earthmate Philosophy lacks a metaphysical foundation and is instead grounded in objective and scientific evidence and emphasizes the fundamental interconnectedness of unity despite plurality. This philosophy draws inspiration from several key philosophical concepts, including Spinoza's "Deus sive Natura", the Absolute in the context of German Idealism, Schelling's Naturphilosophie and considering two philosophical approaches of Collectivism and Organicism in a complex system named the Earth.

As mentioned before, Spinoza's "Deus sive Natura" posits that God and Nature are synonymous, representing a unified substance encompassing all of existence. In addition to this divine unity in existence, UDP also highlights plurality and means that despite the apparent diversity and multitude of things in the world, they are all interconnected and emanate from the same underlying substance or essence. In Spinoza's view, everything is part of a unified whole, and the perceived differences and distinctions are ultimately illusory.

The philosophy of UDP also resonates with German Idealism's concept of the "Absolute" representing the ultimate, all-encompassing reality that integrates all individual concepts and phenomena. In the context of German Idealism, UDP would mean that even though there are countless individual concepts and phenomena, they are all ultimately interconnected and part of the larger whole—the Absolute. It suggests that despite the apparent differences, there is a fundamental interconnectedness between all aspects of reality. This perspective underscores the idea that humans are integral parts of the larger ecological system, advocating for responsible stewardship of the planet.

Schelling's Naturphilosophie, which seeks to understand nature as a unified whole, aligns well with the philosophy of "Earthmate." It emphasizes a dynamic sequence of stages in both organic and inorganic nature and the continuous transformation of nature from one form to another, highlighting the interconnectedness of various functions and forces. This philosophy emphasizes that humans are not separate from nature but are an integral part of it. This implies a mutual interdependence and an ethical responsibility towards maintaining the harmony of the interconnected natural system. This perspective could lead to practical implications, such as advocating for sustainable living practices and a holistic approach to environmental conservation, in line with the UDP and the ideas expressed in Schelling's Naturphilosophie.

In the tapestry of human societies, the philosophy of UDP can stand as a profound testament to our capacity to coexist harmoniously despite our myriad differences. It is an acknowledgment that, while we may hold distinct beliefs, cultures, and perspectives, we share a common destiny that unites us. However, within this intricate balance, there lie potential pitfalls. In the realm of collectivism, there exists a potential for ideological governance to take root. This can manifest as a stifling of individual autonomy and creativity, where the collective ideology becomes paramount, potentially suppressing dissenting voices and stifling innovation. On the flip side, individualism, while celebrating personal freedoms and self-expression, can sometimes veer into the territory of unchecked egoism. This extreme form may lead to a disregard for the Earth, individual needs, and the well-being of a majority of the community, potentially fostering environmental problems, a sense of isolation, and undermining the collective good. Striking a harmonious chord between these forces is an ongoing challenge, one that requires a nuanced understanding of the interplay between individual rights and communal welfare. It is in this delicate dance that societies find their true strength, balancing the tapestry of plurality with the unifying thread of shared destiny of humanity and the earth.

Organicism and Mechanism represent two contrasting paradigms in understanding and approaching biological, ecological, and social systems. The Organic approach emphasizes the interconnectedness, interdependence, and holistic nature of these systems. It recognizes that living organisms and ecosystems are dynamic, self-regulating entities that function as integrated wholes. In biological systems, the Organic perspective acknowledges the complexity of organisms, their adaptability, and their ability to evolve in response to changing environments. Similarly, in ecosystems, it acknowledges the intricate web of relationships among organisms and their environment, highlighting the importance of biodiversity and ecological balance. In social systems, the Organic approach emphasizes the importance of considering cultural, historical, and psychological factors that shape human behavior and interactions. On the other hand, the Mechanical approach tends to view these systems as discrete, predictable entities that can be understood by breaking them down into their constituent parts. While this reductionist approach can be useful in certain contexts, it can limit our understanding of the intricate and nuanced dynamics that exist within biological, ecological, and social systems. Therefore, recognizing the significance of the Organic approach is crucial for a more comprehensive and effective approach to understanding and managing these complex systems.

Therefore, UDP which draws inspiration from a rich tapestry of philosophical ideas, at its core, UDP embraces Spinoza's profound notion of "Deus sive Natura," asserting that the divine unity permeates all of existence. This perspective underscores the intrinsic interconnectedness of all elements in the universe, viewing them as integral parts of a greater whole. Building upon this, UDP incorporates the German Idealism concept of the Absolute, positing an ultimate, all-encompassing reality that harmoniously integrates diverse individual concepts and phenomena. This notion serves as a cornerstone for understanding the underlying unity that unites the manifold aspects of reality. Schelling's Naturphilosophie further riches UDP, illuminating the dynamic procession of stages within both organic and inorganic nature. It elucidates the continuous transformation of natural forms, emphasizing an ongoing and seamless evolution. In the realm of human interaction, UDP navigates the terrain of philosophical approaches, notably Collectivism, which seeks to strike a delicate equilibrium between the forces of collectivism and individualism. UDP recognizes the need for a balanced understanding of how individuals contribute to and coexist within a collective framework, fostering a harmonious societal dynamic. Moreover, UDP embraces the tenets of Organicism, which underscores the intricate, interdependent, and self-regulating nature of complex systems. This perspective highlights the holistic view of interconnectedness, accentuating that the whole transcends the sum of its constituent parts. In this complex system, UDP weaves these philosophical threads into a coherent tapestry, advocating for a world where unity prevails amidst the plurality of existence, recognizing that harmony arises from embracing the dynamic interplay between individual autonomy and collective well-being, all within the larger context of a self-regulating, interconnected whole.

References

Agius, E. (1989). Towards a relational theory of intergenerational ethics. *Bijdragen*, 50(3), 293-313.

Beiser, F. C. (2002). *German Idealism: The Struggle Against Subjectivism*. Harvard University Press.

Briley, D. A., & Wyer Jr, R. S. (2001). Transitory determinants of values and decisions: The utility (or nonutility) of individualism and collectivism in understanding cultural differences. *Social Cognition*, 197–227.

Dudley, W. (2014). *Understanding German Idealism*. Routledge.

Dunham, Jeremy; Grant, Iain Hamilton; Watson, Sean. (2011). *Idealism: A History of a Philosophy. Durham*. McGill-Queen's University Press.

Guilherme, A. (2010). Fichte: Kantian or Spinozian? Three interpretations of the absolute I. *South African Journal of Philosophy = Suid-Afrikaanse Tydskrif vir Wysbegeerte*, 1–16.

Hegel, G. W. (1892). *Hegel's Lectures on the History of Philosophy*. London: Routledge & Kegan Paul.

Hou, X. (2013). *Community capitalism in China: The state, the market, and collectivism*. Cambridge University Press.

Im, T. (2019). *The two sides of Korean administrative culture: Competitiveness or collectivism?* Routledge.

Infantino, L. (2014). *Individualism in modern thought: from Adam Smith to Hayek*. Routledge.

Inwood, M. (1992). *A Hegel Dictionary*. Wiley-Blackwell.

Israel, J. (1996). The Banning of Spinoza's Works in the Dutch Republic (1670-1678). *In Disguised and Overt Spinozism around 1700*, 1-14.

Landau, R. (2013). *The Philosophy of Ibn'Arabi*. Routledge.

Lásźló, É. (2013). *Individualism, collectivism, and political power: A relational analysis of ideological conflict.* Springer.

Limnatis, N. G. (2008). *German Idealism and the Problem of Knowledge:: Kant, Fichte, Schelling, and Hegel.* Springer Science & Business Media.

Macpherson, C. B. (1962). *The political theory of possessive individualism: Hobbes to Locke.* Oxford University Press.

Magnani, L., Aliseda, A., Longo, G., Sinha, C., Thagard, P., & Woods, J. (2015). *Studies in Applied Philosopy, Epistemology and Rational Ethics (Volume 22).* Springer.

McIntyre-Mills, Janet, N. R. A. Romm, and Yvonne Corcoran-Nantes. (2017). *Balancing individualism and collectivism."* Collected papers from special integration group for international systems sciences plus 16 contributors. Springer.

Nadler, S. M. (1999). *Spinoza: A Life.* Cambridge University Press.

Nadler, S. M. (2001). *Spinoza's Heresy: Immortality and the Jewish Mind.* New York: Oxford University Press.

Scruton, R. (2002). *Spinoza: A very short introduction.* USA: Oxford University Press.

Simkins, J. (2014). On the Development of Spinoza's Account of Human Religion. *Intermountain West Journal of Religious Studies*, 5(1), 4.

Solomon, R. C. (1985). *In the spirit of Hegel.* Oxford University Press.

Stewart, M. (2007). *The courtier and the heretic: Leibniz, Spinoza, and the fate of God in the modern world.* WW Norton & Company.

Totaro, P. (2015). *The Young Spinoza: A Metaphysician in the Making.* New York: Oxford University Press.

von Schelling, F. W. (2004). *First Outline of a System of the Philosophy of Nature.* SUNY Press.

Yalom, I. (2012). *The Spinoza problem: A novel.* UK: Hachette.

Zakaras, A. (2022). *The Roots of American Individualism: Political Myth in the Age of Jackson.* Princeton University Press.

3

Earthmate Science

Introduction

Why the studies and efforts tabled in the first season don't work well? We have worked a lot as you see, but why these studies and efforts don't work and we couldn't save the earth effectively? When we confronted with the assertion in the last parts of first season that either the myriad scientific disciplines and intellectual pursuits are ineffective or that the extent of Earth's degradation is substantial, the below reasons prove that the latter is a stark reality supported by compelling statistics and reasoning. The notion that the studies and efforts presented in the initial season are ineffective prompts an exploration into the reasons behind this perceived inadequacy. It is crucial to differentiate between genuine critiques of scientific endeavors and the proliferation of unfounded conspiracy theories.

Sociological and cultural unsustainable development: The sociological and cultural unsustainable development emerges as a significant contributor to the ongoing destruction of Earth, entwined with various factors that shape individual and collective identities. Cultural and social factors wield significant influence, as traditional practices may prove unsustainable or contribute to environmental degradation. Cultures differ not only in their degree of concern but also in the framework and structure of their perspectives on concern (Eisler et al. 2003; Zheng and Yoshino 2003). Addressing these factors requires a profound understanding of local cultures and social dynamics. The lack of public awareness and engagement poses a formidable challenge, given the complexity of environmental issues and the difficulty individuals face in understanding the impacts of their actions. Lack of access to education further exacerbates the problem, hindering communities from participating in efforts to protect the environment or make sustainable choices. Many studies across different countries, such as those by researchers Klineberg et al. (1998), Arcury and Christianson (1993), Ostman and Parker (1987), Hsu and Roth (1996), and Chanda (1999), have observed that individuals with higher levels of education tend to show greater concern for the environment, although some studies, like that of Grendstad and Wollebaek (1998), have found contrasting results. Uneducated individuals, lacking a universal identity approach tied to the Earth, often fail to recognize the urgency of environmental circumstances.

While Smil (1994) rightly argues that the complexity of the connections between population, resources, and the environment defies simplistic a priori generalizations, particularly on a global scale, it remains undeniable that the steady growth of the global population and rapid urbanization significantly intensify resource demand, exerting immense pressure on the environment. While it is acknowledged that urbanization, following the perspective of McDonald et al. (2013), should not be narrowly viewed solely as a problem or solution, it is evident that in many countries it presents environmental challenges such as pollution and habitat loss. Resistance to change is another hurdle, with some individuals and organizations rejecting environmental protection efforts due to disbelief in the science or viewing solutions as threats to their interests. Moreover, despite advances in ecological research, there exists a limited understanding of the Earth's complex ecosystems. This knowledge gap makes it challenging to identify effective solutions to environmental problems, emphasizing the need for continuous research and education.

At the individual level, the three-dimensional identity (personal identity, national identity, and the challenging concept of universal identity) plays a pivotal role. Gifford and Nilsson (2014) consider that personal factors encompass a wide range, including childhood experiences, knowledge and education, personality traits and self-construal, sense of control, values, political and world views, goals, perceived responsibility, cognitive biases, attachment to place, age, gender, and chosen activities. Here the power of states besides highlighted social factors by Gifford and Nilsson (2014) (religion, urban–rural differences, norms, social class, proximity to problematic environmental sites and cultural and ethnic variations) in shaping identity becomes a fundamental issue, emphasizing the geographical significance in contemporary societal dynamics. The root of deepening identity separations and societal divides that can have a destructive effect on the environment, are greed, poverty, jealousy, unacceptable religious ideologies, illogical theories, and reactionary traditions, fostering misunderstanding and conflicts. In fact, here cupidity and avarice or in general greed can play a central role. Public understanding becomes a crucial aspect, with the prevalence of separations stemming from diverse motivations that contribute to the creation and perpetuation of misunderstandings. In essence, the sociological and cultural unsustainable development intertwines with diverse factors, necessitating a comprehensive and culturally sensitive approach to address the root causes of Earth's destruction.

Political and economic unsustainable development: The destructive impact on Earth resulting from political and economic unsustainable development is multifaceted, encompassing various interconnected factors. Although Porter and Linde (1995) suggest dynamic view against static view about environment-competitiveness relationship in industries, we can see that unhealthy competition among the most nations and industries exacerbates the race for economic dominance, often leading to environmentally harmful practices. A major hurdle lies in the lack of political will, as environmental issues become entangled in political agendas. Leaders may hesitate to implement necessary policies, fearing repercussions on economic growth or their own political standing. Even when regulations are in place, their ineffective enforcement permits individuals and companies to engage in detrimental activities without consequence.

Despite the noble aims of environmental protection, global cooperation remains hampered by deep-rooted political, ethnic, and ideological divisions that persist among and within nations (Caldwell and Weiland 1996). Global cooperation is imperative, given the inherently global nature of environmental issues. However, achieving such collaboration faces obstacles like geopolitical tensions, competing priorities, and differing economic development levels among nations or a lack of political will among certain countries. Therefore, although international cooperation is vital, it may be stymied by many obvious and hidden factors.

"Short-term focus", prevalent among decision-makers in both government and the private sector, often results in prioritizing immediate gains over long-term sustainability (Smith and Sharicz 2011). Unsustainable consumption and production patterns, marked by reliance on non-renewable resources and high levels of waste, contribute significantly to environmental degradation. Economic systems and incentives may not align with environmental goals, as a focus on growth and profitability can lead to unsustainable practices or resource exploitation. Conflicting priorities, such as economic growth or national security, compete with environmental concerns, demanding a delicate balance to holistically address challenges.

Rooted in governmental disorders, conflicts arise between national interests and global interests—the interests of the Earth itself. Ideological governments, lacking transparency and employing exclusive language like "we" and "our," may prioritize narrow objectives over collective well-being. Corruption and mismanagement further undermine environmental initiatives, diverting resources or leading to poorly executed projects with harmful consequences.

The role of international organizations becomes crucial in addressing the impunity of governments in harming the Earth and there is often a lack of mechanisms to hold governments accountable for actions that have far-reaching environmental consequences. Urban sprawl and non-standard profit-oriented industries contribute to the exploitation of natural resources and environmental degradation. In essence, the intertwining complexities of political and economic unsustainable development serve as significant contributors to Earth's destruction, demanding comprehensive and coordinated efforts for sustainable solutions.

Scientific and technological unsustainable development: The issue of unsustainable development in the realms of science and technology has emerged as a significant catalyst for the degree of destruction faced by Earth. One of the key challenges lies in the complexity of the environmental problems at hand, which are multifaceted, involving a myriad of factors and stakeholders. Within this context, we would characterize the Earth as an open system, acknowledging the potential presence of unknown relationships and our perceived limitations in comprehensively addressing them, with the boundaries of our Earth reflecting similar constraints dictated by our current knowledge and scientific capabilities (Neugebauer 2006). Effectively addressing these challenges necessitates a profound comprehension of the underlying scientific principles and an openness to collaborative, interdisciplinary approaches. Furthermore, a noteworthy hurdle is the limited scientific consensus on the causes and effects of various environmental problems. This lack of consensus not only complicates the identification of effective solutions but also fuels disagreements over policy and regulatory measures.

A critical aspect contributing to the Earth's degradation is the disparity in understanding between the scientific community and the general populace. Bridging this gap is crucial for garnering public support and fostering informed decision-making on matters that directly impact the environment. This is why Jill Jäger (2006) speaks of a "new contract" between science and society. Moreover, while practices like CSR can have a limited effect in mitigating negative environmental impacts, conflicting interests between major industries and scientific findings exacerbate the problem (Robinson 2012). The pursuit of profit often clashes with environmental preservation efforts, leading to a contentious landscape where economic considerations may supersede ecological concerns. Limited resources also present a formidable challenge to environmental protection and conservation efforts. Adequate funding, scientific expertise, and public support are essential, but scarcity of these resources impedes the implementation of effective solutions.

In addition to these challenges, despite Dietz and Rosa (1994) discussing the potential negative impact of technology through the IPAT model, I believe the inadequacy of technology further hampers sustainable development. The technology required to tackle numerous environmental issues may be nonexistent, economically unviable, or inaccessible to a wide audience. Developing new technologies that are both effective and accessible poses a formidable challenge, requiring concerted efforts in research, development, and dissemination. Until these technological gaps are adequately addressed, the path towards sustainable solutions remains hindered, perpetuating the threats to the Earth's well-being. In essence, the scientific and technological dimensions of unsustainable development contribute significantly to the intricate web of reasons behind the escalating degree of destruction faced by our planet.

The ineffectiveness of studies and efforts tabled in the first season to address Earth's pressing environmental issues can be attributed to a complex and deep interplay of factors spanning scientific, technological, political, economic, sociological, and cultural dimensions. Sociological and cultural unsustainable development by lack of public awareness and engagement, public understanding, lack of access to education, uneducated individuals, the three-dimensional identity or lacking a universal identity, global population's steady growth and rapid urbanization, limited understanding of the Earth's complex ecosystems, greed, poverty, jealousy, unacceptable religious ideologies, illogical theories, and reactionary traditions contribute to misunderstandings and resistance to change. Political and economic unsustainable development exacerbates environmental degradation through unhealthy competition, lack of political will, lack or Ineffective regulations and enforcement, lack or Ineffective global cooperation, short-term focus, unsustainable consumption and production patterns, economic systems and incentives, governmental disorders, Ideological governments, lacking transparency, corruption and mismanagement, lack of mechanisms to hold governments accountable, urban sprawl, profit-oriented industries and conflicting priorities. Scientific and technological unsustainable development hampers progress due to the complexity of environmental problems, limited resources, inadequate technology, limited scientific consensus, and disparities in understanding between the scientific community and the public. These challenges, rooted in greed, geopolitical tensions, and inadequate resources, underscore the need for comprehensive, interdisciplinary solutions. Bridging the gap in understanding, fostering global cooperation, and addressing cultural sensitivities are crucial for effective environmental preservation, demanding unified, concerted efforts to save the Earth from further destruction. of course, these factors, deeply embedded in the historical fabric of human societies, present intricate and enduring challenges that defy simple solutions. The scientific, technological, political, economic, sociological, and cultural dimensions influencing Earth's degradation have evolved over centuries, shaped by complex interactions and historical developments. Greed, geopolitical tensions, and conflicting interests have historical roots, ingrained in the structures of societies and nations. The disparities in understanding, the lack of political will, and resistance to change have deep historical antecedents that resist swift resolution. These multifaceted challenges are not isolated incidents but rather reflections of long-standing patterns ingrained in human behavior and societal structures. Consequently, addressing these issues requires a nuanced and comprehensive understanding of their historical roots, demanding sustained efforts that transcend temporal boundaries and encompass the complexities of our shared human history.

Knowledge and Science

In the realm of science, the methodology is grounded in rigorous testing, peer review, and evidence-based conclusions. However, the field is not immune to criticism, and constructive scrutiny is integral to its progress. Nevertheless, it is crucial to distinguish valid skepticism from baseless conspiracy theories. The scientific process, despite its meticulous nature, is not infallible on a global scale due to the uncontrollable variables and numerous factors at play. Engaging in conspiracy theories within the scientific community is deemed unacceptable as it undermines the credibility of the science. Recognizing the fallibility of scientific methods on a global scale underscores the importance of continuous refinement and adherence to the scientific framework, which, imperfect as it may be, remains the most reliable approach for comprehending the complexities of the world around us.

Similarly, within the realm of science, the positivist methodology has been a cornerstone of knowledge acquisition and progress. This approach fosters a systematic and disciplined exploration of natural phenomena. Nevertheless, even in the scientific community, deeply rooted historical factors can influence perspectives and create challenges. Constructive scrutiny is welcomed, as it contributes to refining scientific understanding. It is essential to uphold the integrity of the scientific discipline by rejecting baseless conspiracies that may arise. It is imperative to recognize that the scientific method, while not flawless, remains the most reliable framework for understanding the world around us.

The foundation of Earthmate's scientific approach rests on a set of simple yet vital principles derived from the scientific method. These principles align seamlessly with the Earthmate philosophy, guiding its commitment to environmental well-being. The Principle of Complex System underscores the recognition of Earth as a multifaceted and interconnected entity, embracing the intricate relationships within its dynamic systems. Emphasizing interconnectedness, the Principle of Systems Thinking guides Earthmate in understanding the holistic dynamics at play. The Principle of Interdisciplinary drives collaboration, acknowledging that environmental issues demand insights from diverse fields. Rooted in long-term thinking, the Principle of Sustainable Development anchors Earthmate's pursuit, advocating for equilibrium among economic, social, and environmental considerations. Recognizing the finite nature of resources, the Principle of Finite Resources steers responsible management. The principle of the circular economy, encompassing the elimination of waste, the continuous use of products and materials, and the regeneration of natural systems, form the foundation for a sustainable and resource-efficient approach to production and consumption. Finally, the Principle of Fragility guides Earthmate in exercising prudent stewardship to protect fragile ecosystems from irreversible harm. These principles collectively form the scientific backbone of Earthmate, shaping its comprehensive and sustainable approach to environmental harmony. Let's dive in to explore further and gain more in-depth information.

Complex System

A complex system refers to a system comprised of numerous interconnected components, such as Earth's global climate, organisms, the human brain, or social and economic organizations. The defining characteristic of complex systems lies in the intricate relationships and interactions between their components, leading to emergent phenomena and behaviors that are difficult to predict. These systems exhibit properties like nonlinearity, emergence, spontaneous order, adaptation, and feedback loops.

The study of complex systems involves investigating how the relationships among a system's segments bring about collective behaviors. This approach stands in contrast to reductionism, which seeks to explain systems by breaking them down into their constituent parts. Reductionists argue that complex systems are often overly complicated, making them difficult to understand and manage, with unpredictable behaviors, challenging modeling and simulation requirements, diminished central control, high expertise needs, and reliance on probabilistic rather than deterministic rules. However, despite these criticisms, complex systems provide powerful insights and tools for understanding and analyzing real-world phenomena that cannot be adequately captured by reductionist approaches. Various disciplines enrich the study of complex systems, including physics, social sciences, mathematics, biology, and economics, making it an interdisciplinary domain. Examples of complex adaptive systems, a special case within complex systems, include ecosystems and the human immune system. Overall, complex systems represent a crucial paradigm in understanding the intricacies and dynamics inherent in various natural and human-made phenomena.

The concept of complex systems has its roots in Warren Weaver's essay (1948) on "Science and Complexity," where he categorized problems based on their level of intricacy. The explicit study of complex systems began to take shape in the 1970s, gaining momentum with the foundation of the Santa Fe Institute (Vemuri 1978; Ledford 2015). Early participants in the institute included distinguished figures such as Murray Gell-Mann, Philip Anderson, Kenneth Arrow, George Cowan, and Herb Anderson (Waldrop 1992). Since then, the interest in complex systems has grown, leading to the establishment of many scientific departments dedicated to the field.

The late 1990s witnessed a surge in the application of mathematical physics methodologies to economic phenomena, giving rise to the field of "econophysics" (Joe et al. 2016). This interdisciplinary approach, rooted in complex systems theory and chaos theory, has spurred a paradigm shift in economic analyses. Notably, Syukuro Manabe, Klaus Hasselmann, and Giorgio Parisi were honored with Nobel Prize for their groundbreaking contributions in understanding complex systems, particularly their contribution to creating more accurate computer models predicting the effects of global warming on Earth's climate.

The Principle of Complex Systems, as articulated by the Earthmate philosophy, aligns seamlessly with the recognition that our world's intricacies cannot be fully understood by focusing solely on discrete topics or objects. The interconnected nature of the world requires an integrative approach, emphasizing the importance of considering the relationships and interactions between various components. This holistic perspective resonates with the study of complex systems, where behaviors emerge from the interplay of multiple factors. The idea that places serve as valuable incubators for integrative thinking (Murphy 2018) mirrors the complex systems approach, emphasizing the need to comprehend how different elements interact within a given environment. In the context of geographic issues, places may become more than just incubators; they may represent complex systems with relationships that extend beyond local control. The Earthmate philosophy's commitment to the Principle of Complex Systems underscores the significance of understanding and managing the complexities inherent in the relationships between environmental, social, and economic factors, contributing to a more comprehensive approach to addressing the challenges that affect the health and well-being of our planet.

Systems Thinking

Systems thinking is a holistic approach to understanding and navigating the complexity of the world. This concept involves examining the interconnectedness and the ripple effect of each component, and relationships within systems rather than focusing solely on individual components. The concept is deeply rooted in the idea of viewing entities as part of larger systems, each with its own set of laws and behaviors. Dana Meadows (2008), in her work "*Thinking in Systems: A Primer*," emphasizes the importance of recognizing systems as integrated wholes composed of diverse, interacting structures and sub-junctions. This framework is particularly valuable for understanding environmental phenomena like anthropogenic HDEs, as explored by Liu et al. (2013) in China's recreation landscapes. Their study, conducted in forest patches of Zhejiang Province, demonstrates how human-induced changes at patch edges trigger ripple effects, disrupting species richness and landscape connectivity far beyond the immediate disturbance zone, affecting both animal- and wind-dispersed species.

The application of systems thinking extends across various domains, from political and economic systems recognized centuries ago to the development of radar systems during World War II. The concept has found relevance in diverse fields, including biology, economics, and social sciences. Systems thinking provides a framework for analyzing and understanding the intricate patterns of behavior that emerge over time within these systems. It serves as a tool for effective action in complex contexts, enabling systemic change. Overall, systems thinking invites a perspective that goes beyond reductionism, encouraging a nuanced understanding of the dynamic interactions within intricate systems.

The roots of systems thinking can be traced back to the seventeenth century, with Robert Hooke's multifaceted use of the concept of system in his contributions to the understanding of planetary motion and the "System of the World." Hooke's exploration of the Ptolemaic system versus the Copernican system highlighted the multifaceted nature of the concept (Marchal 1975). However, it was Isaac Newton's groundbreaking work in the late seventeenth century, particularly in his "Philosophiæ Naturalis Principia Mathematica," that laid the foundation for systems thinking to endure through the ages. Newton's approach, grounded in dynamical systems, continues to influence contemporary applications of systems thinking.

The Earthmate philosophy's integration of the principle of systems thinking into environmental management resonates with the historical development and essence of systems thinking. Recognizing the environment as a complex system composed of interdependent parts, Earthmate philosophy's emphasizes understanding relationships and feedback loops within the system, aligning with the holistic view inherent in systems thinking. The application of systems thinking in environmental management involves considering the entire environmental system, acknowledging dynamic interactions between components and human systems. This approach, akin to the historical evolution of systems thinking, allows for a comprehensive understanding of environmental challenges. The Earthmate philosophy's focus on identifying root causes, such as unsustainable practices and climate change, echoes the systemic perspective inherent in systems thinking. Moreover, the philosophy's call for collaboration among diverse stakeholders mirrors the historical recognition of collective action in systems thinking approaches. Ultimately, by incorporating the principle of systems thinking into environmental management, Earthmate philosophy's aims to address environmental issues holistically, fostering a sustainable and equitable future for both humanity and the planet.

Interdisciplinarity

Interdisciplinarity, in contemporary terms, refers to the amalgamation of multiple academic disciplines into a unified activity or project, particularly in research settings. The essence of interdisciplinarity lies in thinking across traditional boundaries, fostering collaboration between researchers, students, and teachers from different academic schools of thought. It extends beyond the academic realm, finding application in large engineering teams and interdisciplinary fields that transcend traditional boundaries. The term is often associated with educational pedagogies, describing studies that use methods and insights from several established disciplines. Interdisciplinary endeavors become crucial in addressing complex issues like the epidemiology of HIV/AIDS or global warming, where diverse perspectives are essential for comprehensive problem-solving. Overall, interdisciplinarity stands as a dynamic and essential approach to understanding and solving complex problems by integrating knowledge and methodologies from diverse academic disciplines.

The concept of interdisciplinarity has historical antecedents, tracing back to Greek philosophy (Augsburg 2006). The roots of the idea are linked to the notions of unified science, general knowledge, synthesis, and the integration of knowledge (Klein 1990). As history progressed, interdisciplinary approaches were evident in various projects. Seventeenth-century endeavors, such as Leibniz's task to create a system of universal justice, exemplified the need for interdisciplinary collaboration, involving linguistics, economics, management, ethics, law philosophy, politics, and even sinology (Andrés-Gallego 2015). Over time, interdisciplinary programs emerged as a response to perceived neglect or misrepresentation within traditional disciplinary structures, addressing subjects like women's studies or ethnic area studies. The roots of interdisciplinarity lie deep in human intellectual pursuits and the necessity to draw from diverse fields to comprehend complex problems.

The Earthmate philosophy underscores the significance of principle of Interdisciplinary in environmental management, aligning with the broader concept of interdisciplinarity discussed earlier. Recognizing the complexity of environmental challenges like climate change, pollution, and habitat loss, Earthmate philosophy emphasizes collaboration among experts from diverse fields to develop effective solutions. This aligns with the understanding that environmental issues extend beyond scientific dimensions, encompassing social, economic, and political aspects. Principle of Interdisciplinary, as advocated by Earthmate philosophy, involve the integration of knowledge from various disciplines such as ecology, economics, sociology, political science, and engineering. This integration enables a comprehensive understanding of environmental problems and the identification of holistic solutions. For instance, a multidisciplinary team may assess the impacts of a proposed development, considering ecological, social, and economic implications. Moreover,

Earthmate philosophy's emphasis on engaging local communities aligns with the broader view of interdisciplinarity, recognizing the importance of incorporating diverse perspectives in the decision-making process. By fostering collaboration among experts and involving local communities, interdisciplinary approaches to environmental management, as embraced by Earthmate philosophy, contribute to the development of solutions that are not only effective and sustainable but also equitable, paving the way for a more resilient and harmonious future for the planet.

Sustainable Development

The concept of sustainable development has a rich history rooted in the evolution of societal attitudes towards natural resource management and environmental conservation. It traces its origins to the seventeenth and eighteenth centuries in Europe when ideas about sustainable forest management emerged (Blewitt 2014). To be specific, John Evelyn and Hans Carl von Carlowitz laid the groundwork for the sustainable use of natural resources, emphasizing the importance of planting trees and managing forests for sustained yield. This concept gained momentum in response to the depletion of timber resources and the understanding that overexploitation could lead to environmental degradation. The core idea, as defined in the Brundtland Report of 1987, is to meet the needs of the present without compromising the ability of future generations to meet their own needs (WCED 1987). In the contemporary context, sustainable development addresses pressing global challenges such as climate change, biodiversity loss, and social inequality. The formalization of sustainable development occurred during the 1992 Earth Summit in Rio de Janeiro, where the Rio Process was initiated. The term "sustainability" is often viewed as a long-term goal, while "sustainable development" refers to the processes and pathways undertaken to achieve this goal (UNESCO 2012). UNESCO distinguishes between the two by emphasizing that sustainability is an overarching objective, while sustainable development involves the practical steps taken to reach that objective. UNGA further solidified the concept in 2015 with the adoption of

SDGs for the period 2015 to 2030. The SDGs provide a comprehensive framework addressing global challenges such as poverty, inequality, climate change, and environmental degradation. The adoption of the SDGs reflects a concerted global effort to guide countries toward a future where economic prosperity, social equity, and environmental integrity coexist.

Sustainable development serves as an organizing principle that seeks to harmonize human development goals with the capacity of natural systems to provide essential resources and ecosystem services. The concept involves finding a delicate balance between economic development, environmental protection, and social well-being. The three pillars of sustainable development—economic, environmental, and social—highlight the interconnectedness of these dimensions. Wolfram Mauser (2006), in the book "Earth System Science in the Anthropocene" edited by Ehlers and Krafft (2006), argue that sustainable development is the only viable long-term strategy for addressing the impacts of global change. However, the concept has faced criticism, with some questioning its feasibility and others calling for more concrete actions to address the complex and intertwined issues of our time. For instance, in a great work, Holden et al. (2014) argued that the lack of a clear definition for sustainable development has led to its misuse in addressing local issues rather than global challenges, and proposed a return to the Brundtland Report's original concept, with specific thresholds for ecological sustainability, basic needs, and equity that, with the aid of technology and behavioral changes, could be met by 2030. Over time, sustainable development has evolved beyond its initial focus on intergenerational equity to encompass socially inclusive and environmentally sustainable economic growth.

The Principle of Sustainable Development, as articulated in the Brundtland Report, underscores the need to meet present needs without compromising the ability of future generations to meet their own needs. This principle resonates with the Earthmate philosophy, which encourages a holistic approach to living in harmony with the Earth. Both emphasize the importance of responsible resource management, considering the limitations imposed by technology and social organization on the environment's capacity. The Earthmate philosophy, rooted in ecological awareness and ethical choices, mirrors the call for a balance between economic development, environmental protection, and social well-being inherent in sustainable development. By intertwining these principles, there emerges a shared ethos that seeks to guide humanity towards a future where human needs are met without jeopardizing the delicate balance of the planet's ecosystems, fostering a symbiotic relationship between people and the Earth.

Scarcity and Finite Resources

Scarcity, as defined by Paul Samuelson and William Nordhaus, implies that in a world of limited resources, choosing one option inevitably requires sacrificing another (Samuelson and Nordhaus 2010). Scarcity, in this context, signifies the gap between limited resources and theoretically limitless human wants. It is a driving force in economic theory, influencing how societies allocate resources and make choices. Heterodox economists believe that scarcity is not solely a consequence of resource limits but also a result of human activity and social provisioning (Daoud 2010). The concept extends to absolute and relative scarcity, with absolute scarcity linked to the Malthusian idea of resource constraints leading to population-related crises. On the other hand, according to Lionel Robbins, relative scarcity is a core concept in economics, signifying that the available resources are not enough to produce everything people want. In the modern world, scarcity necessitates competition for resources, leading to trade-offs and the allocation of scarce resources through mechanisms such as the price system (Heyne et al. 2013). Understanding scarcity is essential for comprehending economic decision-making and resource allocation in societies.

The concept of scarcity has deep roots in economic thought, dating back to influential economists like Lionel Robbins and Thomas Robert Malthus. In 1932, Robbins defined economics as the study of human behavior in the face of limited resources, emphasizing the relationship between ends and scarce means with alternative uses. Malthus, in the late eighteenth century, laid the theoretical foundation for understanding scarcity with his observations on population growth and resource limitations. He introduced the idea of absolute scarcity, where population growth outpaces the availability of resources, leading to a Malthusian catastrophe. The concept evolved with distinctions between absolute and relative scarcity. Absolute scarcity, as explained by economist Daly, refers to the scarcity of resources in general, intensifying as population and per-capita consumption approach the biosphere's carrying capacity (Daoud 2010). On the other hand, relative scarcity, as emphasized by modern economic theory, highlights the insufficient resources to produce all desired goods, forming the basis for economic choices.

The Earthmate philosophy aligns seamlessly with the principle of scarcity in economics, emphasizing the finite nature of resources as a fundamental tenet. Sustainability, a core concept in Earthmate philosophy, underscores the imperative to use Earth's resources in a manner that fulfills present needs without jeopardizing the ability of future generations to meet their own. This philosophy acknowledges that the Earth's resources are limited and that human activities exert significant impacts on ecosystems and natural resources. The recognition of the finite nature of resources echoes the economic principle of scarcity, where the availability of resources is constrained. The Earthmate philosophy calls for a shift towards sustainable resource use, advocating for renewable energy sources, efficient resource utilization, and the preservation and restoration of ecosystems. This harmonious approach mirrors the economic imperative to address scarcity through thoughtful resource allocation and utilization. By embracing sustainability and acknowledging the limitations of Earth's resources, the Earthmate philosophy paves the way for a more sustainable and resilient future, aligning with the broader ethos of responsible resource management.

Circularity or Circular Economy (CE)

While some, including Murray et al. (2017) and Korhonen et al. (2018), attribute the circular economy to China's economic development strategies (featured in both the 11th and 12th 'Five Year Plans'), as well as to the EU, several national governments, and numerous global businesses, the concept itself cannot be pinpointed to a single date or author. Instead, it has roots in a variety of intellectual traditions and schools of thought. The idea can be linked to industrial ecology, biomimicry, and cradle-to-cradle design principles. Industrial ecology, focusing on material and energy flows through industrial systems, laid the foundation for the circular economy. The notion of emulating nature's patterns and strategies, known as biomimicry, and the holistic approach of cradle-to-cradle design principles also contributed to the development of the circular economy. According to Kirchherr et al. (2017), Geissdoerfer et al. (2017) and Korhonen et al. (2018), while there are still ambiguities surrounding the concept, a circular economy represents a transformative model of resource production and consumption. While the circular economy is seen as a pathway to sustainable development, critics argue about the feasibility of achieving zero waste and question its social and environmental benefits. Recognizing the fallacy lies in the acknowledgment that while absolute zero waste might be challenging; significant strides can still be made towards minimizing and managing waste in a more sustainable manner. Despite debates, the circular economy remains a key concept in addressing pressing environmental challenges and fostering sustainable economic practices. It aimed at addressing global challenges such as climate change, biodiversity loss, and pollution. At its core, the circular economy seeks to create a closed-loop system, minimizing resource inputs, waste, pollution, and carbon emissions. It diverges from the linear economy's "take, make, waste" approach by promoting a structure based on 10 common circular economy strategies (i.e. recover, recycling, repurpose, remanufacture, refurbish, repair, re-use, reduce, rethink, refuse) existing materials and products for as long as possible (Morseletto 2020). The Ellen MacArthur Foundation played a pivotal role in promoting the circular economy in Europe and the Americas. The Foundation claims that the three fundamental principles underpinning the circular economy are designing out waste and pollution, keeping products and materials in use, and regenerating natural systems. Business models can play a crucial role in facilitating this transition, with approaches like product-as-a-service and sharing platforms optimizing resource utilization. Despite all debates, circular economy strategies extend beyond individual products to encompass entire industries, cities, and even countries.

Rachel Carson's watershed work, *Silent Spring* (2002), was first published in1962, after being serialized in The New Yorker magazine. The pivotal notes and the book illuminate the harmful effects of man-made pesticides on the environment, marking a crucial moment in the emerging environmental movement and highlighting the detrimental relationship between unchecked economic growth and environmental degradation. This growing awareness significantly influenced economist Kenneth Boulding, who, in 1966, addressed these concerns in his seminal work *The Economics of the Coming Spaceship Earth*, as discussed in the collection of essays in the book *Environmental Quality in a Growing Economy* (Jarrett 1966). Boulding contrasted the exploitative "cowboy economy"—which thrives on the endless consumption of resources—with the sustainable "spaceship economy," which recognizes the necessity of a circular economic system that operates within the Earth's ecological limits and finite resources to ensure long-term sustainability. Additionally, numerous influential figures, such as Allan Kneese from the United States, Walter Stahel from Switzerland, Hiroshi Komiyama from Japan, and Ma Jun from China, have been pivotal in advancing the concept of the 'circular economy.' Their groundbreaking work has underscored the critical importance of resource efficiency and environmental sustainability, driving innovative approaches to waste reduction and resource management across various sectors and regions. Their collective efforts highlight a transformative shift towards a regenerative economic model that aims to harmonize economic growth with ecological stewardship. The concept of the circular economy gained significant momentum in the early 2000s, particularly as China began integrating it into its industrial and environmental policies. Notably, the circular economy was emphasized in both the 11th and 12th Five-Year Plans, and the state-owned China Coal industry adopted it as a guiding philosophy, among other initiatives. The European Union formally introduced its vision in 2014, with a new CEAP launched in March 2020. According to the claim of the French Ministry of Ecological Transition and Territorial Cohesion, major events in the global economic landscape, such as increases in raw material prices, Chinese control over rare earth materials, and the 2008 economic crisis, further accelerated the adoption of circular economy principles.

The foundational principles of the circular economy, centered on eliminating waste and pollution or mitigating of them, prolonging the use of products and materials, and revitalizing natural systems, harmoniously resonate with the principles of the Earthmate philosophy. Earthmate philosophy emphasizes a holistic and sustainable approach to living in harmony with the environment, acknowledging the interconnectedness of all living beings. The circular economy's core tenets resonate with the Earthmate philosophy by promoting responsible resource consumption, waste reduction, and the restoration of natural systems. Both concepts advocate for a shift from the linear "take, make, waste" model to a regenerative system that mirrors the resilience and balance observed in nature. Embracing circular economy principles within the Earthmate philosophy means recognizing the finite nature of resources, valuing ecosystem health, and striving for a harmonious coexistence between humanity and the planet. Together, these frameworks offer a comprehensive vision for a sustainable future that prioritizes ecological well-being, minimizes environmental impact, and fosters a balanced relationship between human activities and the Earth's intricate ecosystems.

Environmental Fragility

Environmental fragility refers to the sensitivity and susceptibility of ecosystems, landscapes, or regions to disturbances and changes, both natural and anthropogenic. This concept encompasses a broad range of environmental factors, including biodiversity loss, soil degradation, water scarcity, and climate vulnerability. Nilsson and Grelsson (1995), Margules (1986), and Smith and Theberge (1986) propose that fragility can be defined as the inverse of stability, meaning that higher fragility corresponds to lower stability, and vice versa. Environments exhibit diverse levels of resilience and adaptability within the context of environmental fragility, making them subject to analysis and evaluation. In the praxis of governance, more fragile environments are more prone to negative impacts from human activities, making them crucial focal points for conservation efforts and sustainable resource management. Understanding environmental fragility is essential for developing strategies to mitigate the adverse effects of human-induced changes and climate variability, fostering a more resilient and sustainable relationship between humanity and the natural world.

The concept of environmental fragility has its roots in the growing awareness of the Earth's vulnerability to human activities and natural processes. As environmental science emerged as a distinct field in the mid-twentieth century, researchers and policymakers began to recognize the intricate balance within ecosystems and the delicate nature of the planet's environmental systems. The idea gained prominence as scientific evidence pointed towards the impact of human activities such as deforestation, pollution, and climate change on the Earth's resilience. The realization that certain ecosystems are particularly susceptible to disturbances and disruptions led to the conceptualization of environmental fragility. Over time, this concept has become a crucial component in understanding the susceptibility of different regions and ecosystems to environmental changes and the need for sustainable practices to protect our planet.

In the realm of environmental issues, the Principle of Environmental Fragility underscores the varying capacities of ecosystems to withstand and recover from disturbances, emphasizing the need for tailored approaches to conservation and sustainable management. This principle aligns with the Earthmate philosophy, which advocates for responsible stewardship of the planet by recognizing the interconnectedness of human activities and the natural environment. Essential tools in assessing environmental fragility include the Ecological Footprint, which measures human impact on ecosystems, and Acceptance Capacity, which gauges an ecosystem's ability to absorb stress without crossing critical thresholds. These metrics aid in evaluating the sustainability of human practices and guide the formulation of policies that align with the delicate balance of Earth's ecological systems. Embracing the Earthmate philosophy alongside the principles of environmental fragility provides a holistic framework for addressing environmental challenges and fostering a harmonious coexistence between humanity and the natural world.

Scientific Actions

In confronting both present and future environmental challenges, the Earthmate philosophy advocates for a strategic amalgamation of three fundamental principles of Scientific Actions. Firstly, the Prevention of environmental problems emerges as a crucial guiding principle, urging a proactive stance in curbing the root causes of environmental issues. This involves aligning human activities with sustainable practices, acknowledging the finite nature of resources, and understanding the fragility of ecosystems. Secondly, the Conservation and restoration of ecosystems and biodiversity play a pivotal role, emphasizing the need to safeguard and revitalize natural habitats. By adhering to principles of biodiversity, resilience, and conservation, Earthmate endeavors to preserve the intricate web of life on Earth, recognizing its inherent value. Lastly, the philosophy underscores the importance of Adapting to and building resilience against environmental changes. In acknowledging the inevitability of shifts in the Earth's systems, the Earthmate philosophy aligns with principles of resilience urging the adoption of adaptive strategies to mitigate and navigate the impacts of changing environmental conditions. By integrating these three principles, Earthmate endeavors to forge a holistic and sustainable path toward addressing the intricate web of environmental challenges we face today and in the future. Let's dive in to explore further and gain more in-depth information.

Prevention of Environmental Problems

The concept of prevention in the context of environmental issues involves taking proactive measures to avoid or minimize the occurrence of problems that could harm the planet. This approach recognizes that it is often more effective and efficient to address potential environmental threats before they escalate into crises. Prevention entails a wide range of actions, including education for children and citizens, empowering individuals actively involved in the realm of ecomedia, reducing pollution, minimizing waste, conserving resources, and advocating for sustainable policies and practices. By focusing on prevention, the goal is to create a resilient and harmonious relationship between human activities and the natural environment.

The importance of prevention in environmental stewardship cannot be overstated. Rather than reacting to the consequences of environmental degradation, prevention aims to address the root causes and mitigate the impact on ecosystems and biodiversity. This necessitates a heightened focus on education and further research. Not only does prevention safeguard the planet's health, but it also contributes to long-term sustainability, ensuring that future generations inherit a world that is thriving and capable of meeting their needs. The cost-effectiveness of prevention further underscores its significance, as investing in sustainable practices and policies early on can save resources that would otherwise be expended in addressing and rectifying environmental damage.

In the Earthmate philosophy, prevention holds a central place. This philosophy recognizes that fostering a deep connection with the environment is crucial for creating a sustainable future. By embracing the Earthmate philosophy, individuals, communities, and societies align their values and actions with the goal of preventing environmental harm. This philosophy encourages a holistic approach that integrates scientific research, policy development, and individual responsibility. By promoting a collective commitment to preventing environmental problems, the Earthmate philosophy aims to inspire a global movement towards a more sustainable and ecologically balanced world.

Conservation and Restoration of Ecosystems and Biodiversity

The concept of conservation and restoration within the framework of the Earthmate philosophy revolves around preserving and reviving the intricate web of life on Earth. Biodiversity, encompassing genetic, species, and ecosystem diversity, lies at the core of this philosophy. Recognizing the interdependence of all living beings, the Earthmate philosophy emphasizes the importance of safeguarding and rehabilitating ecosystems. This involves not only protecting individual species but also maintaining the balance and functioning of entire ecosystems, crucial for the provision of essential services such as air and water purification, pollination, and nutrient cycling.

The significance of conservation and restoration efforts cannot be overstated, considering the profound impact human activities have had on biodiversity and ecosystem health. The Earthmate philosophy underscores the urgent need to address these issues to safeguard the health and well-being of both humanity and the planet. As habitats disappear and species face extinction, the philosophy advocates for proactive measures to ensure the long-term sustainability of ecosystems and the services they provide, highlighting the interconnectedness of environmental health with the overall prosperity of the Earth and its inhabitants.

In aligning with the Earthmate philosophy, conservation and restoration initiatives become not just environmental imperatives but integral components of social and economic well-being. These efforts offer opportunities for economic benefits, such as sustainable tourism and recreation, and promote the responsible use of natural resources. Furthermore, they contribute to the preservation of cultural values intertwined with nature. By emphasizing the link between environmental conservation, economic prosperity, and cultural richness, the Earthmate philosophy provides a holistic approach to building a sustainable and resilient future for the planet and its diverse inhabitants.

Adapting to and Building Resilience Against Environmental Changes

Adapting to and building resilience against environmental changes form a fundamental pillar of the Earthmate philosophy, a holistic approach that underscores the interconnectedness of human and ecological well-being. In its essence, adaptation involves adjusting to evolving environmental conditions, while resilience focuses on the ability of ecosystems and communities to withstand shocks and recover swiftly from disturbances. With the Earth's ecosystems constantly in flux and human-induced factors exacerbating these changes, addressing challenges such as extreme climate change, pollution, and deforestation requires a proactive stance that aligns with the Earthmate philosophy.

The importance of adapting and building resilience lies in mitigating the adverse effects of environmental pressures on both natural and human systems. The Earthmate philosophy recognizes that resilience is not only about safeguarding biodiversity but also about promoting the overall health and functionality of ecosystems. By bolstering adaptive capacity and managing environmental risks, the philosophy aims to create a robust foundation for sustainable development. This includes adopting conservation practices, sustainable land use, and adaptive management strategies, along with enhancing infrastructure resilience and emphasizing preparedness and response planning.

The Earthmate philosophy's emphasis on adaptation and resilience extends beyond ecological considerations, recognizing their pivotal role in ensuring social and economic well-being. Resilient communities and ecosystems are better equipped to navigate environmental challenges, fostering a sense of stability and continuity. By integrating resilience into its core tenets, the Earthmate philosophy offers a comprehensive roadmap towards a sustainable and equitable future, acknowledging the intricate interplay between human societies and the natural world. It champions a harmonious coexistence where adaptability and resilience become cornerstones for a more balanced and resilient planet.

Conclusion

In conclusion, the challenges faced in addressing Earth's pressing environmental issues are deeply rooted in a complex interplay of factors across scientific, technological, political, economic, sociological, and cultural dimensions. Sociological and cultural unsustainable development, coupled with political and economic complexities, has hindered effective solutions. These challenges are embedded in historical human behaviors and societal structures, resisting simple solutions due to historical roots, geopolitical tensions, and conflicting interests.

The Earthmate philosophy, grounded in key scientific principles, offers a holistic approach to these challenges. It aligns with the principles of Complex Systems, Systems Thinking, and Interdisciplinary collaboration, recognizing Earth's multifaceted and interconnected nature.

Through an interdisciplinary approach and systems thinking, Earthmate aims to holistically address environmental challenges, considering the dynamic relationships within ecological, social, and economic systems. This philosophy aligns with historical developments in science and systems thinking, emphasizing collaboration, understanding relationships, and a comprehensive approach to address the challenges affecting our planet's health and well-being.

In essence, the call to action involves bridging the gap in understanding, fostering global cooperation, and addressing cultural sensitivities. A unified, concerted effort is needed to save the Earth from further destruction. The Earthmate philosophy, anchored in scientific principles and interdisciplinary collaboration, provides a foundation for such efforts, aiming for a sustainable and harmonious future for both humanity and the planet.

The Earthmate philosophy, aligned with sustainable development principles, emphasizes the delicate balance between economic prosperity, social well-being, and environmental protection. It echoes the Principle of Sustainable Development, advocating for responsible resource management and a holistic approach to living in harmony with the Earth.

Finite resources, a common thread in both economics and the Earthmate philosophy, acknowledge the fundamental reality of scarcity. Emphasizing the importance of understanding and addressing scarcity, both perspectives guide economic decision-making and ensure the sustainable use of Earth's limited resources.

The Earthmate philosophy also embraces circular economy principles, promoting a transformative model that minimizes waste, pollution, and carbon emissions. This aligns with its emphasis on responsible resource consumption, waste reduction, and the restoration of natural systems, fostering a regenerative approach to production and consumption.

Environmental fragility, central to the Earthmate philosophy, underscores the sensitivity of ecosystems to disturbances. It aligns with the philosophy's call for responsible stewardship of the planet, recognizing the interconnectedness of human activities and the natural environment.

Scientific Actions, advocated by the Earthmate philosophy, involve a strategic amalgamation of prevention, conservation and restoration, and adaptation to address environmental challenges. These principles form a cohesive strategy for sustainable and resilient living, fostering a harmonious coexistence between human activities and the Earth's intricate ecosystems.

Ultimately, the Earthmate philosophy envisions a world where individuals, communities, and societies collectively commit to preventing environmental problems, conserving and restoring ecosystems, and adapting to environmental changes. By embracing this philosophy, humanity can strive towards a harmonious coexistence with the planet, fostering sustainability, resilience, and a balanced relationship between human activities and the Earth's intricate ecosystems. As we navigate the complex issues of our time, the Earthmate philosophy offers a guiding ethos for a more sustainable and ecologically balanced future.

References

Andrés-Gallego, J. (2015). Are humanism and mixed methods related? Leibniz's universal (Chinese) dream. *Journal of Mixed Methods Research*, 9(2), 118-132.

Arcury, T. A., & Christianson, E. H. (1993). Rural-urban differences in environmental knowledge and actions. *The Journal of Environmental Education*, 25(1), 19-25.

Augsburg, T. (2006). *Becoming interdisciplinary: An introduction to interdisciplinary studies*. New York: Kendall Hunt Publishing.

Blewitt, J. (2014). *Understanding sustainable development (2nd ed.)*. London: Routledge.

Caldwell, L. K., & Weiland, P. S. (1996). *International environmental policy: from the twentieth to the twenty-first century*. Duke University Press.

Carson, R. (2002). *Silent spring*. New York: Mariner Books.

Chanda, R. (1999). Correlates and dimensions of environmental quality concern among residents of an African subtropical city: Gaborone, Botswana. *The Journal of Environmental Education*, 30(2), 31–39.

Daoud, A. (2010). Robbins and Malthus on scarcity, abundance, and sufficiency: The missing sociocultural element. *American Journal of Economics and Sociology*, 69(4), 1206-1229.

Dietz, T., & Rosa, E. A. (1994). Rethinking the environmental impacts of population, affluence and technology. *Human ecology review*, 1(2), 277-300.

Ehlers, E., & Krafft, T. (2006). *Earth system science in the anthropocene*. Bonn: Springer.

Eisler, A. D., Eisler, H., & Yoshida, M. (2003). Perception of human ecology: cross-cultural and gender comparisons. *Journal of Environmental Psychology*, 23(1), 89-101.

Geissdoerfer, M., Savaget, P., Bocken, N. M., & Hultink, E. J. (2017). The Circular Economy–A new sustainability paradigm? *Journal of cleaner production*, 143, 757-768.

Gifford, R., & Nilsson, A. (2014). Personal and social factors that influence pro-environmental concern and behaviour: A review. *International journal of psychology*, 49(3), 141-157.

Grendstad, G., & Wollebaek, D. (1998). Greener still? An empirical examination of Eckersley's ecocentric approach. *Environment and behavior*, 30(5), 653-675.

Heyne, P. T., Boettke, P. J., & Prychitko, D. L. (2013). *The economic way of thinking (13th ed.)*. Pearson.

Holden, E., Linnerud, K., & Banister, D. (2014). Sustainable development: Our common future revisited. *Global environmental change*, 26, 130-139.

Hsu, S. J., & Roth, R. E. (1996). An assessment of environmental knowledge and attitudes held by community leaders in the Hualien area of Taiwan. *The Journal of Environmental Education*, 28(1), 24-31.

Jäger, J. (2006). Sustainability Science. In E. Ehlers, & T. Krafft, *Earth system science in the anthropocene* (pp. 19-26). Bonn: Springer.

Jarrett, H. (1966). *Environmental quality in a growing economy: Essays from the sixth RFF Forum*. Baltimore, MD: Johns Hopkins Press.

Joe, H. Y., Ruiz Estrada, M. A., & Yap, S. F. (2016). The evolution of complex systems theory and the advancement of econophysics methods in the study of stock markets crashes. *Labuan Bulletin of International Business & Finance (LBIBf)*, 14(1), 68-83.

Kirchherr, J., Reike, D., & Hekkert, M. (2017). Conceptualizing the circular economy: An analysis of 114 definitions. *Resources, conservation and recycling*, 127, 221-232.

Klein, J. T. (1990). *Interdisciplinarity: History, theory, and practice*. Detroit: Wayne state university press.

Klineberg, S. L., McKeever, M., & Rothenbach, B. (1998). Demographic predictors of environmental concern: It does make a difference how it's measured. *Social science quarterly*, 734–753.

Korhonen, J., Honkasalo, A., & Seppälä, J. (2018). Circular economy: the concept and its limitations. *Ecological economics*, 143, 37-46.

Ledford, H. (2015). How to solve the world's biggest problems. *Nature*, 525 (7569): 308–311.

Liu, B., Su, J., Chen, J., Cui, G., & Ma, J. (2013). Anthropogenic halo disturbances alter landscape and plant richness: A ripple effect. *PLoS One*, 8(2), e56109.

Marchal, J. H. (1975). On the Concept of a System. *Philosophy of Science*, 42(4), 448-468.

Margules, C. R. (1986). Wildlife conservation evaluation. In M. B. Usher, *Conservation evaluation in practice* (pp. 297-314). London: Chapman and Hall.

Mauser, W. (2006). Global Change Research in the Anthropocene: Introductory Remarks. In E. Ehlers, & T. Krafft, *Earth System Science in the Anthropocene* (pp. 3-4). Bonn: Springer.

McDonald, R. I., Marcotullio, P. J., & Güneralp, B. (2013). *Urbanization and global trends in biodiversity and ecosystem services. Urbanization, biodiversity and ecosystem services: challenges and opportunities: a global assessment.* Dordrecht, Heidelberg, New York, and London: Springer.

Meadows, D. H. (2008). *Thinking in systems: A primer.* chelsea green publishing.

Morseletto, P. (2020). Targets for a circular economy. *Resources, conservation and recycling*, 153, 104553.

Murphy, A. B. (2018). *Geography: Why it matters.* Cambridge: Polity Press.

Murray, A., Skene, K., & Haynes, K. (2017). The circular economy: an interdisciplinary exploration of the concept and application in a global context. *Journal of business ethics*, 140, 369-380.

Neugebauer, H. J. (2006). What about Complexity of Earth Systems? In E. Ehlers, & T. Krafft, *Earth system science in the anthropocene* (pp. 27-38). Bonn: Springer.

Nilsson, C., & Grelsson, G. (1995). The fragility of ecosystems: a review. *Journal of Applied Ecology*, 677–692.

Ostman, R. E., & Parker, J. L. (1987). Impact of education, age, newspapers, and television on environmental knowledge, concerns, and behaviors. *The Journal of Environmental Education*, 19(1), 3-9.

Porter, M. E., & Linde, C. V. D. (1995). Toward a new conception of the environment-competitiveness relationship. *Journal of economic perspectives*, 9(4), 97-118.

Robinson, J. G. (2012). Common and conflicting interests in the engagements between conservation organizations and corporations. *Conservation biology*, 26(6), 967-977.

Samuelson, P. A., & Nordhaus, W. D. (2010). *Economics (19th ed.)*. New York: McGraw-Hill.

Smil, V. (1994). How many people can the earth feed? *Population and Development Review*, 255–292.

Smith, P. A., & Sharicz, C. (2011). The shift needed for sustainability. *The learning organization*, 18(1), 73-86.

Smith, P.G.R. & Theberge, J.B. (1986). A review of criteria for evaluating natural areas. *Environmental Management*, (10),715-734.

UNESCO. (2012). *Education for sustainable development*. Paris: UNESCO.

Vemuri, V. (1978). *Modeling of complex systems: an introduction (Operations research and industrial engineering)*. New York: Academic Press.

Waldrop, M. M. (1992). *Complexity: The emerging science at the edge of order and chaos*. New York: Simon and Schuster.

WCED. (1987). *Our common future*. Oxford: Oxford University Press.

Weaver, W. (1948). Science and complexity. *American scientist*, 36(4), 536-544.

Zheng, Y., & Yoshino, R. (2003). Diversity patterns of attitudes toward nature and environment in Japan, USA, and European nations. *Behaviormetrika*, 30, 21-37.

4

Earthmate Ethics

Introduction

Ethics is a branch of philosophy that deals with the study of moral principles and the concept of right and wrong conduct. It explores questions related to how individuals and societies should behave and make decisions. Ethics provides a framework for evaluating the morality of actions and determining what is considered morally acceptable or unacceptable. While morality often refers to personal beliefs and principles, ethics often involves broader considerations and systematic reflections on moral values. A significant challenge within the realm of ethics and even morality lies in their potential disregard for personal interests and even logical interests. Several philosophical, cultural, religious, and ethical viewpoints, including skeptics, Moral Relativists, and Ethical Nihilists, cast doubt on the legitimacy of ethics, prompting inquiries into the existence of any objective foundation for moral values. Moreover, individuals with a secular orientation may question the authenticity of moral principles derived from religious sources on non-environmental ethics that such principles may indeed carry legitimacy. However, recent scientific discoveries, some of which were unveiled in the last season, challenge these skeptical perspectives on environmental ethics. These findings suggest that numerous philosophical traditions and moral frameworks offer profound insights into objective and external truth and reality. Despite the skepticism expressed by some, there is mounting evidence supporting the existence of a solid foundation for ethics. Some of the key points about the negative environmental effects of them include:

Resource Exploitation: Skepticism and Moral Relativism may contribute to a lack of consensus on ethical practices, potentially leading to increased resource exploitation and environmental degradation and promotion of Consumerist cultures, influenced by hedonistic desires. One such value, termed "deep anthropocentrism," involves the deliberate separation of human society from the natural world, dismissing ecology as a valid ethical consideration, and shaping the criteria for what skeptics deem as legitimate knowledge (Jacques 2006). They can accelerate climate change, such as deforestation and excessive use of fossil fuels. The constant pursuit of pleasure and material possessions, characteristic of hedonistic and consumerist cultures, may divert attention away from environmental concerns, reducing motivation for collective action on climate change.

Consumerism: The constant pursuit of pleasure and material gratification inherent in consumerism, often driven by hedonistic tendencies, results in significant environmental consequences. This overconsumption pattern leads to environmental degradation, depletion of finite natural resources, and the generation of substantial waste. Overconsumption, influenced by hedonistic desires for more, becomes a driving force behind resource exploitation, contributing to activities harmful to biodiversity, such as overfishing, deforestation, and pollution. Additionally, in urban areas, the hedonistic-driven culture of consumerism exacerbates environmental challenges related to water, energy, and waste management through resource overconsumption. The production and disposal of single-use plastics, propelled by the hedonistic desires embedded in consumerist cultures, contribute to pervasive plastic pollution, harming marine life and ecosystems. There is a widely accepted consensus in the energy sector that affluent nations, comprising one-fifth of the global population yet consuming approximately two-thirds of all fossil fuels, need to stabilize their energy consumption over the next generation, primarily through significant efficiency improvements, to facilitate relatively swift and affordable consumption growth in poorer countries (WEC 1993). While the pace and scope of specific changes may vary, there is a clear imperative for a fundamental, enduring shift away from prevailing consumption patterns towards a more sustainable way of life, encompassing stabilized and ultimately reduced energy usage and revised food consumption as critical components of this overdue transition from mindless consumerism and a borrowing economy in the Western context (Smil 1994). The interconnectedness of hedonistic-driven consumerism with resource exploitation and environmental degradation underscores the importance of addressing the underlying philosophical perspectives that downplay the importance of ethics in environmental stewardship.

Short-Term Focus: Hedonistic tendencies, often associated with consumerism, may prioritize short-term pleasures over long-term considerations, contributing to a lack of commitment to sustainable practices that address climate change. On the other hand, many large and small companies, as well as governments, often pursue higher returns through short-term focus strategies mostly called early return plans. While these strategies may appear profitable and logical in the short run, they frequently prove to be detrimental in the long term and to society as a whole. Although some researchers question the direct link between intense commercial and industrial competition and environmental degradation—like Porter and Linde (1995), who argues that the infusion of innovation in many businesses challenges the notion that heightened competition necessarily harms the environment—there is no denying that traditional approaches to commercial and industrial activities still dominate a significant portion of national economies. These conventional practices, often focused on maximizing short-term gains, continue to exert considerable pressure on environmental resources, despite the growing presence of innovative, eco-friendly strategies. By prioritizing immediate gains over sustainable investments, these entities contribute to the depletion of natural resources, increased pollution, and the exacerbation of climate change. Such approaches, which emphasize short-term productivity and quick profits rather than environmentally friendly and sustainable practices, ultimately threaten not only the environment but also the global economy. The long-term costs of these actions can be immense, leading to significant environmental degradation, economic instability, and a burden on communities worldwide.

Disregarding Intergenerational Ethics: Philosophical perspectives that question the authenticity of ethics may lead to a lack of consideration for intergenerational ethics, also called obligations to future generations, resulting in insufficient efforts to address climate change for the benefit of future generations. In this field, Agius (1989) proposes a relational ethical theory grounded in process thought, redefining humanity, nature, and God from an interdependent standpoint, to argue that unborn generations have rights that impose responsibilities on current generations, emphasizing collective responsibility and social justice across time. He advances this ethical theory with a Metaphysical Basis, but considering that its focus will be on specific communities, this approach may not be necessary. Instead, according to the Earthmate Philosophy, it could be more effectively developed using objective facts as its foundation.

Climate Policy Obstruction and Inaction: Skepticism toward the authenticity of ethics can hinder the development and implementation of robust climate policies, slowing down efforts to mitigate the impacts of climate change. Moral Relativism may lead to a lack of universal ethical standards for addressing climate change, resulting in inconsistent or delayed global cooperation on mitigation strategies. Moreover, Skepticism and Moral Relativism may foster reluctance to adopt sustainable practices that could improve air quality, such as transitioning to cleaner energy sources and reducing emissions.

Policy Stagnation: Ethical Nihilism can result in a lack of commitment to environmental policies and regulations, hindering progress in addressing pressing environmental issues.

Lack of Collective Responsibility: Skepticism and Moral Relativism may foster a harmful sense of individualism and utilitarianism, reducing the sense of collective responsibility for environmental conservation efforts. Ethical Nihilists also may reject the moral responsibility to address climate change or question the moral imperative to address air quality issues, contributing to a lack of motivation for collective action, accountability, and accountability for environmental stewardship. When examining the issue of collective responsibility in the context of climate change, one can consider various perspectives, similar to the two approaches discussed by Säde Hormio (2023). The first approach views climate change as an inherently collective problem arising from aggregated individual actions, while the second focuses on the responsibility of collective entities, such as states and corporations. If collective entities fail to adopt effective institutional and legal approaches to address this issue, the ethical dimension will become a central issue because if ethics, alongside awareness, becomes an intrinsic force for each individual, it is valuable for addressing the problem. However, it is not necessarily reliable or sufficient factor on its own, as it lacks the comprehensive impact and coordination that institutional and collective actions can provide.

Delay in Sustainable Practices: Doubts about the authenticity of ethics, as expressed by Ethical Nihilists, may lead to delays in adopting sustainable practices, hindering the transition to environmentally friendly alternatives. It also may question the moral imperative for sustainable urban development, potentially hindering efforts to create environmentally friendly and resilient cities.

Undermining Environmental Advocacy: Doubts about the authenticity of moral values, as expressed by skeptics, may undermine the advocacy for urgent action on climate change, clean air initiatives, sustainable urbanization, the protection of biodiversity, and the protection of oceans, delaying public and political mobilization. Skepticism also may lead to a lack of support for stringent environmental regulations that aim to control air pollution, resulting in compromised air quality, as individuals may question the moral imperative to protect the environment.

Inconsistent Ethical Standards: Moral Relativism can give rise to divergent ethical norms among different cultures, potentially introducing inconsistencies in environmental protection endeavors. This variability may manifest in differing ethical viewpoints regarding climate-related matters, the significance of air quality, the value placed on biodiversity, the importance of ocean health, and the repercussions of urbanization and population growth. Such disparities impede the establishment of a cohesive global strategy to address climate change, air pollution, conservation efforts, marine protection, and sustainable urban development.

Diminishing Focus on Environmental Education: Skepticism regarding the authenticity of moral values has the potential to lessen the importance placed on environmental education. This reduction in emphasis may result in decreased awareness and understanding of crucial aspects, including the significance of ecological balance, the ethical imperative to address climate change, the ethical imperative to tackle air quality concerns, the ethical necessity of balancing urbanization with ecological preservation, the ethical obligation to protect biodiversity, and the ethical imperative to safeguard ocean health.

Disregard for Interconnectedness: Philosophical perspectives that downplay the significance of moral values may contribute to a disregard for the interconnectedness of ecosystems, potentially leading to unintended environmental consequences.

Inadequate Environmental Policies: Skepticism and Moral Relativism can hinder the development and implementation of robust environmental policies, leaving ecosystems vulnerable to exploitation and degradation. Philosophical perspectives that downplay the importance of ethics in environmental matters may lead to insufficient action on climate change, indirectly affecting air quality and biodiversity through altered atmospheric conditions, habitats, and ecosystems.

Impaired Environmental Activism: Ethical Nihilism's rejection of intrinsic moral value may lead to a lack of motivation for individuals to engage in environmental activism, impeding collective efforts to address environmental challenges.

Habitat Destruction: Skepticism and Moral Relativism may lead to a lack of consensus on the importance of biodiversity, potentially resulting in insufficient efforts to protect habitats from destruction by various human activities. Ethical Nihilists also may question the moral imperative to preserve biodiversity and ocean health, potentially contributing to a lack of motivation for conservation efforts and policies.

Disregarding Ecosystem Services: Skepticism and Moral Relativism may result in a failure to recognize the vital ecosystem services provided by biodiversity, such as pollination, water purification, and climate regulation, and of course oceans, such as climate regulation, nutrient cycling, and habitat support.

Overfishing: Skepticism and Moral Relativism may contribute to a lack of ethical guidelines for sustainable fishing practices, leading to overfishing and depletion of marine resources.

Urban Sprawl: Skepticism and Moral Relativism may lead to a lack of ethical guidelines for urban planning, contributing to uncontrolled urban sprawl and increased environmental footprint. It also may result in a failure to consider the ecological impact of urbanization, leading to the destruction of natural habitats and a decline in biodiversity.

Disregarding Respiratory Health: Skepticism and Moral Relativism may result in a failure to recognize the health implications of poor air quality, leading to respiratory issues and other health problems for urban populations.

Industrial Emissions: Consumerist cultures, influenced by hedonistic desires, may drive industrial activities that emit pollutants, contributing to poor air quality and posing health risks to communities.

We have discussed the negative aspects of certain philosophical, intellectual, and ethical tendencies, such as skepticism, moral relativism, ethical nihilism, and utilitarianism, in the context of environmental conservation. However, it is important to recognize that some of these perspectives may also play a reformative and positive role, particularly in challenging traditional and superstitious beliefs. By questioning outdated norms, these approaches can contribute to a more critical and informed understanding of our responsibilities toward the environment and the Earth, ultimately fostering more sustainable practices. The Earthmate Ethics are grounded in the belief that human beings have a responsibility to act as stewards of the Earth, and to act in a way that ensures the long-term health and well-being of the planet and all its inhabitants. Here I explore and suggest the ethical principles of being an Earthmate. Some of the key ethical principles that underlie the Earthmate Ethics include:

Respect for Nature

Respect is a fundamental aspect of human interaction and social dynamics, encompassing a deep appreciation for the inherent worth, dignity, and rights of individuals. It involves treating others with courtesy, consideration, and esteem, regardless of differences in opinions, beliefs, or backgrounds. Respect for nature is beyond social interactions and refers to a deeper inner sense and an ethical and responsible approach to the environment, recognizing the intrinsic value of the natural world and understanding the interconnectedness between humans and the broader ecosystem. It involves acknowledging the importance of preserving and sustaining the environment for current and future generations. Respect for nature in fact is respect for future generations and one of the crucial principles of intergenerational ethics than needs to think beyond our own interests. The most profound philosophical approach to this principle is encapsulated in Paul W. Taylor's work, particularly in *Respect for Nature: A Theory of Environmental Ethics* (1986, 2001), where he introduces the theory of biocentrism—a worldview distinct from, yet related to, deep ecology. This perspective, standing in opposition to anthropocentrism, aligns with the anti-speciesism movements that emerged in the wake of masculism and racism. At the heart of the Earthmate Ethics lies a deep respect for nature and all living things, and that human beings have a responsibility to respect and protect the natural world. This respect is rooted in the belief that all beings have inherent value and deserve to be treated with dignity and care. Here are key functional aspects of respect for nature.

Sustainable Practices: This respect for nature and all living things manifests itself in a number of ways within the Earthmate Ethics. Adopting sustainable practices in daily life and industry is one of them that is a manifestation of respect for nature and encourages us to adopt sustainable practices that minimize harm to the environment and other species. It also emphasizes the importance of protecting and preserving biodiversity, recognizing that each species has a unique role to play in the larger web of life. This includes using resources efficiently, reducing waste, and supporting environmentally friendly initiatives.

Environmental Education: Promoting awareness and understanding of environmental issues is a key component of respect for nature. Educating oneself and others about the importance of sustainable living and conservation helps foster a sense of responsibility. Environmental education is not confined to any specific age, method, or setting; it can begin in early childhood within the family or in preschool within the educational system and continue throughout a person's life on Earth, as will be discussed in detail in the next chapter. One of the relevant articles on Respect for Nature in the context of preschool education in Sweden—a pioneering country in this field—is authored by Ärlemalm-Hagsér (2013), who compellingly demonstrates that years after the non-binding action plan of the United Nations for sustainable development, namely Agenda 21, global education on nature conservation and, at a higher level, respect for nature, remains far from satisfactory.

Conservation: Respecting nature involves efforts to conserve and protect ecosystems, wildlife, and natural habitats. This may include supporting conservation organizations, participating in conservation projects, actively working to conserve natural resources, protect biodiversity, and minimize human impact on ecosystems, and advocating for policies that promote environmental protection.

Ethical Consumption: Being mindful of the environmental impact of personal consumption choices is a way to show respect for nature. This can involve choosing products with minimal ecological footprints, supporting eco-friendly businesses, and reducing reliance on single-use items.

Understanding Interconnectedness: Recognizing the interconnectedness of all living beings and ecosystems is fundamental to respecting nature. Actions that harm one part of an ecosystem can create ripple effects throughout the entire system, as discussed in Chap. 3 on systems thinking and referencing the work of Liu et al. (2013). Thus, the Earthmate Ethics promotes the idea of coexistence and interconnectedness, recognizing that all beings are part of a larger ecosystem and that our actions can set off chain reactions throughout the system. It encourages us to think beyond our own interests and to consider the well-being of other species and ecosystems in our decision-making.

Preservation of Biodiversity: Respecting nature includes valuing and working to preserve the rich diversity of plant and animal species. The Earthmate Ethics recognizes that humans are not the only beings that inhabit the Earth, and that our actions can have profound impacts on other species and ecosystems. Thus, we have a responsibility to act with compassion and empathy towards all living things, and to consider their needs and interests in our decision-making processes. This kind of respecting nature involves efforts to prevent species extinction and protect ecosystems that support diverse forms of life.

Reducing Environmental Footprint: Respecting nature involves minimizing the environmental footprint of human activities. This can be achieved through energy conservation, waste reduction, and adopting eco-friendly technologies and practices.

Ultimately, the respect for nature and all living things that lies at the heart of the Earthmate Ethics is a recognition of our shared responsibility to care for the planet and all its inhabitants. By practicing respect for nature, individuals and societies contribute to the well-being of the planet and promote a sustainable and harmonious relationship between humans and the environment. It is a call to action to live in harmony with the natural world, recognizing that we are but one part of a larger web of life, and that our actions have consequences that extend far beyond ourselves. This approach recognizes that a healthy and balanced natural world is essential for the overall health and prosperity of the Earth and its inhabitants.

Interdependence

Interdependence refers to a mutual reliance and interconnectedness between two or more entities, where they depend on each other for support, cooperation, or mutual benefits. It emphasizes the idea that entities are not isolated or self-sufficient but instead rely on each other in various ways to function and thrive. The interdependence of all living things refers to the interconnected relationships and mutual dependencies that exist within ecosystems. In nature, various species and organisms are interdependent, meaning they rely on each other in different ways for survival, growth, and reproduction. This interdependence of all living things is a fundamental principle of the Earthmate Ethics, and it underpins my approach to environmental management and conservation. We recognize that we are part of a larger ecosystem, and that our actions can have ripple effects throughout the system. Thus, we have a responsibility to act in ways that are mindful of these interconnections and to consider the well-being of other species and ecosystems in our decision-making. This interdependence is a fundamental aspect of ecological systems and plays a crucial role in maintaining the balance of life on Earth. Here are some key aspects of the interdependence of all living things.

Food Chains and Webs: In ecosystems, food chains and webs represent one of the most evident forms of interdependence, where organisms are linked through predator–prey relationships, and one organism serves as a source of food for another. Disruptions in these relationships can have cascading effects throughout the ecosystem. Ingram and Brklacich (2006) describe GECAFS as a new interdisciplinary research approach that addresses the complex interactions between GEC and food provision and food security, aiming to develop adaptive strategies and foster dialogue among policymakers, donors, and scientists to support policy formulation and societal well-being. This approach underscores the importance of understanding how disruptions in food systems due to GEC can impact both ecological stability and human food security, highlighting the need for effective adaptation strategies.

Symbiotic Relationships: Many species engage in symbiotic relationships, where two or more organisms live in close association with each other. This can take various forms, such as mutualism (both species benefit), commensalism (one benefits, and the other is unaffected), and parasitism (one benefits at the expense of the other). All these symbiotic relationships serve as a profound reminder of the intricate interdependence between all animal and plant species within Earth's ecosystem, highlighting the delicate balance that sustains life on the planet.

Pollination and Seed Dispersal: Plants often depend on animals for pollination and seed dispersal. Pollinators, such as bees and butterflies, facilitate the reproduction of flowering plants by transferring pollen between flowers. Similarly, animals play a role in dispersing seeds, helping plants colonize new areas. Landim et al. (2024), along with many other similar scientists, demonstrate that pollination and seed dispersal are vital for ecosystem health, and their proposed framework underscores that developing targeted strategies to reintroduce species—by evaluating their unique interactions and resource needs—is crucial for effectively restoring these essential ecological functions. Awareness of these events, along with many other similar insights, provides us with a deep understanding of the importance of interdependence and biodiversity.

Nutrient Cycling: Decomposers, such as bacteria and fungi, play a crucial role in breaking down organic matter, returning nutrients to the soil. This nutrient cycling is essential for the growth of plants, which, in turn, provide food for herbivores and other organisms. While we often overlook or fail to fully appreciate many events that highlight the interdependence of human life alongside other species within ecosystems, numerous phenomena, such as nutrient cycling discussed in this section, remain entirely invisible to us.

Climate Regulation: Plants, trees, forests, grasslands, wetlands, mangroves, soil, algae, phytoplankton, oceans, and seas all contribute to climate regulation by absorbing CO_2 and releasing O_2 through photosynthesis or carbon sequestration processes. Oceans, seas, and soil serve as significant carbon sinks, while wetlands and mangroves store large amounts of carbon in their biomass and soils, playing a critical role in reducing GHGs and mitigating climate change. This process is essential for maintaining the balance of gases in the atmosphere and supporting the respiration of many organisms.

Ecosystem Stability: The interdependence of species contributes to the overall stability of ecosystems. Biodiversity, or the variety of life, enhances an ecosystem's resilience to disturbances. A diverse array of species provides different functions and services, making the ecosystem more adaptable to environmental changes.

Understanding and appreciating the interdependence of all living things is crucial for conservation efforts and sustainable management of ecosystems. This understanding is rooted in the recognition that every species has a unique role to play in the larger web of life, and that our actions can have profound impacts on other species and ecosystems.

The Earthmate Ethics recognizes that all living things are interconnected and interdependent, and that the health of the planet and all its inhabitants is dependent on the health of its ecosystems. Moreover, the Earthmate Ethics recognizes that human well-being is intimately linked to the health of the natural world. We depend on clean air, water, and soil for our survival, and we rely on other species for food, medicine, and other resources. Disruptions or imbalances in these interconnected relationships can have far-reaching consequences for the health and functioning of the entire ecosystem. Thus, the health of the natural world is not only important for its own sake, but also for our own well-being.

At its core, the Earthmate Ethics advocates for living in harmony with nature, acknowledging the intricate interdependence of all life forms and the essential role of ecosystems and biodiversity in maintaining the planet's health. Respecting this interconnectedness underscores the philosophy's essence, recognizing our role as stewards of the planet— preserving its resources and ecosystems for the prosperity of future generations. This recognition of the interdependence of all living things is a call to action to live in harmony with the natural world, recognizing that our fates are intertwined with the fates of other species and ecosystems. It is a reminder that we are but one part of a larger web of life, and that our actions have consequences that extend far beyond ourselves. By embracing this principle of interdependence and acting in ways that honor our interconnectedness with the natural world, we can work towards a more just, equitable, and sustainable future for all.

Responsibility and Sustainability

Responsibility refers to the state or quality of being accountable for one's obligations, decisions, and actions which involves recognizing and accepting the consequences of one's choices and behaviors. Responsibility encompasses various aspects of life, including personal, professional, social, economic, environmental and ethical dimensions. This multifaceted concept involves a sense of duty, accountability, and the willingness to act in a manner that considers the consequences of one's actions on oneself and others. The Stockholm Conference cemented an expanded concept of national environmental responsibility with far-reaching implications for the future of international political, legal, and organizational frameworks addressing environmental challenges (Caldwell and Weiland 1996). This aligns closely with Earthmate Ethics, which asserts that individuals must act as stewards of the planet, taking accountability for the environmental and social impacts of their actions.

Sustainability also refers to the ability to meet the needs of the present without compromising the ability of future generations to meet their own needs. It encompasses balancing economic, social, and environmental considerations to ensure that development and progress are not achieved at the expense of the planet's health or the well-being of current and future generations.

There are three main pillars of sustainability:

1. Environmental Sustainability: This aspect focuses on preserving and protecting the natural resources and ecosystems that support life on Earth.
2. Social Sustainability: Social sustainability is concerned with promoting social well-being, equity, and justice and includes aspects such as community development, social inclusion, access to education and healthcare, and the protection of human rights.
3. Economic Sustainability: Economic sustainability involves creating a stable promoting fair trade, responsible consumption and production, and economic practices of thriving economy that do not deplete resources or exploit communities and provides opportunities for current and future generations.

As discussed in various research works across diverse fields and approaches, such as those by Kruger et al. (2018), Sáez de Cámara et al. (2021), Alshuwaikhat and Abubakar (2008), Urbaniec et al. (2017), Filho et al. (2015), Psarommatis and Bravos (2022), Samadhiya and Agrawal (2020), and Asif et al. (2008), as well as many other works, achieving sustainability requires a holistic and integrated approach that considers the interconnections between these three pillars. Many organizations, businesses, and governments are adopting sustainable practices and policies that reflect this holistic mindset, aiming to address challenges like climate change, resource depletion, and social inequality. The Earthmate Ethics embodies this approach, recognizing the responsibility of humanity to act as stewards of the Earth, ensuring that their actions positively impact both the environment and society. By promoting sustainable development, The Earthmate Ethics seeks to create a more equitable world while safeguarding the long-term health of the planet.

Sustainability and responsibility form the cornerstone of the Earthmate Ethics. Central to the Ethics is the acknowledgment of humanity's responsibility to safeguard the planet, ensuring its health and viability for the generations to come. This responsibility necessitates making deliberate choices in how we lead our lives, conduct our work, and utilize resources, considering the enduring impact on the environment and all living entities. The Earthmate Ethics emphasizes that sustainability transcends environmental concerns alone, encompassing social and economic dimensions. The interconnectedness of the planet's well-being with that of its inhabitants underscores the imperative for responsible and collaborative efforts in forging a sustainable future that benefits everyone. This collaborative ethos extends to decision-making processes regarding resource utilization, infrastructure development, and economic growth, emphasizing the needs of all living entities. Embracing sustainability and responsibility as fundamental values, the Earthmate Ethics envisions a world where all living entities can thrive both presently and, in the years, to come.

Equity and Justice

Equity and justice are two fundamental concepts in the realms of law, ethics, and social systems. While they are closely related, they carry distinct meanings. Equity involves fairness in ensuring individuals or groups receive what is just and fair, accounting for their specific needs, circumstances, and differences. It seeks to address disparities and provide individuals with what they need to achieve a level playing field. On the other hand, justice is a broader concept encompassing fairness, morality, and the administration of the law. It means ensuring that everyone is treated in line with ethical standards and legal principles.

In this context, there are various types of justice. Distributive justice deals with the fair distribution of resources and opportunities, aiming to reduce social and economic inequalities. Retributive justice focuses on punishing wrongdoers to restore balance, while restorative justice emphasizes repairing harm caused and restoring relationships between affected parties.

In addition to these forms of justice, procedural justice plays a critical role in ensuring fairness in decision-making processes. It involves the right to participate in decisions that affect one's life, ensuring that all individuals, especially marginalized and vulnerable groups, have a voice in shaping policies and actions that impact them. In the context of environmental sustainability, procedural justice ensures that communities affected by climate change or resource management decisions are actively involved in the process, promoting transparency, accountability, and inclusivity. This is particularly crucial when addressing environmental inequities, as decisions made without input from those most affected can perpetuate disparities and undermine efforts towards sustainable development (Thomas and Twyman 2005).

The Earthmate Ethics weaves these principles into the fabric of sustainable environmental management. In this framework, equity calls for the fair distribution of natural resources and environmental benefits, ensuring that marginalized and vulnerable communities, who often bear the brunt of environmental degradation, receive fair treatment and access to resources. Meanwhile, environmental justice emphasizes recognizing the rights of all living beings and safeguarding the planet for future generations.

Therefore, The Earthmate Ethics advocates for policies and actions that address these environmental disparities. It calls for engaging and empowering affected communities, ensuring their voices are heard in decision-making processes, and providing the necessary support to overcome environmental challenges. The concept underscores the collective responsibility required for sustainability, emphasizing collaboration between individuals, organizations, and governments to create a more just and equitable society that prioritizes the health and well-being of the planet and its inhabitants.

Ultimately, The Earthmate Ethics envisions a future where responsible actions and sustainable choices lead to equity and justice for all, ensuring the planet's well-being and the fair treatment of its inhabitants for generations to come.

Collaboration, Tolerance and Shared Goals

Collaboration, tolerance, and shared goals are concepts related to working together effectively in various settings, such as teams, organizations, or communities. Collaboration is the act of working with others to achieve a common goal or objective. It encompasses individuals or groups actively participating, sharing ideas, resources, and responsibilities to achieve a collective outcome. Its key elements are positive communication (open and effective communication), mutual respect (valuing collaborators and perspectives, and fostering a positive and supportive working environment), and shared responsibilities (distributing abilities, and responsibilities based on individual strengths and expertise).

Tolerance refers to the ability to accept and respect differences among individuals, whether those differences are related to opinions, backgrounds, cultures, or other factors. In a collaborative setting, tolerance helps create an inclusive environment where diverse perspectives are valued. Its key elements are open-mindedness (Tolerant individuals are open to considering and understanding different viewpoints, even if they differ from their own.), respect for diversity (Tolerance involves recognizing and appreciating the diversity of backgrounds, experiences, and perspectives that team members bring to the collaboration.), and conflict resolution (Tolerance also plays a role in managing conflicts constructively by promoting understanding and finding common ground.).

Shared goals are common objectives that a group of individuals or organizations collectively work toward. These goals provide a unifying purpose for collaboration and help align efforts toward a common outcome. Its key elements are clarity (shared goals should be clearly defined, understood, and agreed upon by all collaborators to ensure everyone is working towards the same objectives.), alignment (Individual and team efforts should be aligned with the overarching shared goals to maximize efficiency and effectiveness.), and adaptability (As circumstances change, collaborators may need to adapt and modify their goals to stay relevant and responsive to evolving needs.). All in all, collaboration involves working together, tolerance emphasizes accepting and respecting differences, and shared goals provide a common purpose for collective efforts. Together, these concepts contribute to successful and harmonious group dynamics in various settings.

The Earthmate Ethics underscores the necessity of collaboration, tolerance, and shared goals in addressing complex environmental challenges. Recognizing that sustainable environmental management requires concerted efforts across sectors and disciplines, the Earthmate Ethics advocates for a commitment to working together despite differences towards common objectives. In the face of multifaceted environmental issues, collaboration and tolerance become essential, as diverse individuals and groups bring varied perspectives, experiences, and priorities to the table. This collective approach fosters a more comprehensive understanding of problems and facilitates the development of effective, inclusive solutions. Emphasizing shared goals is crucial, providing a unifying purpose and direction for environmental management efforts, aligning resources toward achieving sustainable outcomes such as reducing GHG emissions, conserving biodiversity, and promoting renewable energy sources. These principles of collaboration, tolerance, and shared goals not only advance environmental sustainability but also embody the Earthmate Ethics' commitment to inclusivity, diversity, and shared responsibility for a just and equitable society.

Conclusion

In conclusion, this chapter: "Earthmate Ethics" delved into the foundational principles guiding the Earthmate philosophy's ethical framework. Ethics, encompassing the study of moral principles and the distinction between right and wrong conduct, forms the basis for understanding how individuals and societies should behave and make decisions. Despite skepticism and relativism challenging the legitimacy of ethics, recent scientific findings support the existence of a solid foundation for ethical values, especially in environmental contexts. The chapter identifies key negative environmental effects stemming from ethical skepticism, moral relativism, and ethical nihilism, including resource exploitation, overconsumption, short-term focus, and policy stagnation. The Earthmate philosophy's ethical principles counter these challenges by emphasizing respect for nature, interdependence, responsibility, sustainability, equity, and justice. By promoting collaboration, tolerance, and shared goals, the Earthmate Ethics envisions a harmonious and sustainable future, where individuals work together despite differences to achieve common environmental objectives. The philosophy recognizes the interconnectedness of all living things and calls for responsible, equitable, and just actions to ensure the well-being of the planet and its inhabitants for current and future generations.

References

Agius, E. (1989). Towards a relational theory of intergenerational ethics. *Bijdragen*, 50(3), 293-313.
Alshuwaikhat, H. M., & Abubakar, I. (2008). An integrated approach to achieving campus sustainability: assessment of the current campus environmental management practices. *Journal of cleaner production*, 16(16), 1777-1785.
Ärlemalm-Hagsér, E. (2013). Respect for Nature--A Prescription for Developing Environmental Awareness in Preschool. *Center for Educational Policy Studies Journal*, 3(1), 25-44.
Asif, M., de Bruijn, E. J., Fisscher, O. A., & Steenhuis, H. J. (2008, September). Achieving sustainability three dimensionally. *Proceedings of the 4th IEEE International Conference on Management of Innovation and Technology (ICMIT)* (pp. 423–428). Piscataway, New Jersey, USA: Institute of Electrical and Electronics Engineers (IEEE).
Caldwell, L. K., & Weiland, P. S. (1996). *International environmental policy: from the twentieth to the twenty-first century*. Duke University Press.
Filho, W. L., Shiel, C., & Paço, A. D. (2015). Integrative approaches to environmental sustainability at universities: an overview of challenges and priorities. *Journal of Integrative Environmental Sciences*, 12(1), 1-14.
Hormio, S. (2023). Collective responsibility for climate change. *WIREs Climate Change*, 14(4), e830.
Ingram, J., & Brklacich, M. (2006). Global Environmental Change and Food Systems GECAFS: A new interdisciplinary research project. In E. Ehlers, & T. Krafft, *Earth System Science in the Anthropocene* (pp. 217-228). Bonn: Springer.

Jacques, P. (2006). The rearguard of modernity: Environmental skepticism as a struggle of citizenship. *Global Environmental Politics*, 6(1), 76-101.

Kruger, C.; Caiado, R. G. G.; França, S. L. B.; & Quelhas, O. L. G. (2018). A holistic model integrating value co-creation methodologies towards the sustainable development. *Journal of Cleaner Production*, 191, 400-416.

Landim, A. R.; Guimarães Jr, P. R.; Fernandez, F. A.; & Dias, A. T. C. (2024). A framework for the restoration of seed dispersal and pollination. *Restoration Ecology*, e14151.

Liu, B., Su, J., Chen, J., Cui, G., & Ma, J. (2013). Anthropogenic halo disturbances alter landscape and plant richness: A ripple effect. *PLoS One*, 8(2), e56109.

Porter, M. E., & Linde, C. V. D. (1995). Toward a new conception of the environment-competitiveness relationship. *Journal of economic perspectives*, 9(4), 97-118.

Psarommatis, F., & Bravos, G. (2022). A holistic approach for achieving sustainable manufacturing using zero defect manufacturing: A conceptual framework. *Procedia CIRP* (pp. 107, 107–112). Amsterdam, Netherlands: Elsevier.

Sáez de Cámara, E., Fernández, I., & Castillo-Eguskitza, N. (2021). A holistic approach to integrate and evaluate sustainable development in higher education: The case study of the University of the Basque Country. *Sustainability*, 13(1), 392.

Samadhiya, A., & Agrawal, R. (2020, March). Achieving sustainability through holistic maintenance-key for competitiveness. *10th IEOM Annual International Conference* (pp. 10–12). Dubai, United Arab Emirates (UAE): IEOM Society International.

Smil, V. (1994). How many people can the earth feed? *Population and Development Review*, 255–292.

Taylor, P. W. (1986). *Respect for Nature: A Theory of Environmental Ethics*. Princeton, NJ: Princeton University Press.

Taylor, P. W. (2001). *Respect for Nature: A Theory of Environmental Ethics (25th anniversary ed.)*. Princeton, NJ: Princeton University Press.

Thomas, D. S., & Twyman, C. (2005). Equity and justice in climate change adaptation amongst natural-resource-dependent societies. *Global environmental change*, 15(2), 115-124.

Urbaniec, K., Mikulčić, H., Rosen, M. A., & Duić, N. (2017). A holistic approach to sustainable development of energy, water and environment systems. *Journal of cleaner production*, 155, 1-11.

WEC. (1993). *Energy for Tomorrow's World*. New York: St. Martin's Press.

5

Earthmate in Action

Introduction

The Earthmate principles serve as a guiding framework for fostering sustainable practices and environmental stewardship in our rapidly changing world. Comprising the Earthmate Philosophy, Earthmate Science, and Earthmate Ethics, these principles draw inspiration from various ancient and modern philosophies and sciences to advocate for unity amidst diversity, interdisciplinary collaboration, and responsible stewardship of the planet. Within this framework, we delve into three distinct sections, each focusing on the roles, practices, and challenges encountered in adopting Earthmate principles across a spectrum of stakeholders and various sectors.

Section 1: The Role of Individuals, Communities, and Organizations

In this section, we explore the diverse roles played by individuals, communities, and organizations in embracing Earthmate principles. From ordinary citizens to influential figures, each group contributes to shaping environmental attitudes and behaviors. The nine categories under scrutiny encompass a wide array of stakeholders, including environmental activists, educational institutions, governmental entities, and profit-oriented industries. Through examining both positive and negative contributions, I aim to elucidate the complexities inherent in fostering widespread environmental consciousness and action.

Section 2: Successful Practices and Initiatives

Here, I turn our attention to successful endeavors driving the implementation of Earthmate principles. Educational institutions, consultative bodies, and research institutions lead the charge in fostering environmental literacy and sustainable practices. Similarly, non-profit organizations, governmental entities, and international bodies play pivotal roles in spearheading conservation efforts and policy advancements. Across the nine categories, we highlight innovative partnerships, collaborative initiatives, and best practices aimed at mitigating environmental degradation and promoting resilience in the face of global challenges.

Section 3: Challenges and Obstacles

In the final section, we confront the challenges hindering the widespread adoption of Earthmate principles. Regulatory frameworks, cultural norms, economic incentives, and technological limitations present formidable barriers to progress across various sectors. By dissecting these obstacles within each category, we seek to identify actionable solutions and pathways forward. From advocating for policy reforms to fostering cross-sectoral cooperation, addressing these challenges requires a concerted effort from all stakeholders involved.

As we navigate through these discussions, let us remain cognizant of the urgent need for collective action and shared responsibility in safeguarding our planet for future generations. By leveraging the principles of Earthmate philosophy, science, and ethics, we can chart a course towards a more sustainable and resilient future, where unity, compassion, and stewardship guide our interactions with the natural world. Join us in this journey of exploration and discovery as we strive to create a world where humanity and the environment thrive in harmony.

The Role of Individuals, Communities, and Organizations

Individuals

Ordinary People

Ordinary individuals wield significant influence in adopting the Earthmate principles by embodying the Earthmate philosophy (UDP), the Earthmate Science, and the Earthmate Ethics in their daily lives as catalysts for positive change. On the affirmative side, their conscientious choices and behaviors in eco-friendly lifestyles through small yet impactful actions, individuals contribute to the interconnectedness of humanity and nature advocated by the Earthmate principles. These conscientious choices and behaviors serve as potent instruments in embodying the Earthmate principles, and these individuals, acting as the Earthmate ambassadors through their daily decisions and lifestyles, demonstrate an acute awareness of the interconnectedness among all living entities and the environment. These Earthmate ambassadors not only make sustainable choices contributing to a healthier planet but also cultivate a healthy lifestyle for themselves and others, creating a symbiotic relationship between personal well-being and a healthy environment. When ordinary people seamlessly incorporate the Earthmate principles into their routines, they become active agents of positive transformation, fostering a heightened sense of responsibility toward the environment and championing sustainable practices. By embracing sustainable practices, such as minimalism, zero waste living, community-supported agriculture, the preference for eco-friendly products, conserving energy, and supporting local ecosystems and initiatives, ordinary people demonstrate a commitment to responsible stewardship of the planet. These actions align with Earthmate's emphasis on the delicate balance between individual autonomy and collective well-being, fostering a sense of interconnectedness and shared responsibility among communities worldwide.

Another positive dimension lies in empowering individuals to contribute to a global movement for environmental consciousness. The true potential for change emerges when individual actions coalesce into a collective societal movement. Aligned with the Earthmate principles, individuals can actively participate in the Earthmate's mission to promote unity amid the diverse fabric of existence, fostering a grassroots movement with far-reaching impacts on both local and global ecosystems. This positive contribution is marked by individuals who embrace the Earthmate principles, adding to the emergence of a collective consciousness that acknowledges the intricate interconnectedness of all life forms. From adopting eco-friendly lifestyles to actively participating in local clean-up networks, these ordinary endeavors contribute to the overarching vision of a harmonious coexistence with nature. The philosophy's recognition of the dynamic interplay between individual autonomy and collective well-being finds resonance in grassroots movements led by everyday citizens dedicated to effecting positive environmental change.

On the flip side, challenges surface in the form of resistance to change and a reluctance to adopt a mindset prioritizing unity over individualism. Overcoming ingrained habits and societal norms that perpetuate environmental degradation becomes a substantial obstacle, necessitating targeted educational campaigns and community engagement. In conclusion, ordinary people may assume a dual role in the Earthmate principles' adoption and may be potential obstacles in shaping the environmental narrative. On one hand, they may serve as the driving force behind successful practices and initiatives, embodying the Earthmate principles in their daily lives. On the other hand, they may act conversely. The most challenges arise from resistance to change and disparities in awareness and resources. The Earthmate principles, however, acknowledges the importance of addressing these challenges and emphasizes the necessity for a nuanced understanding of the interplay between individual actions and collective well-being. The widespread adoption of the Earthmate principles by ordinary individuals is ultimately imperative for forging a harmonious and sustainable world.

Experts and Academia

Experts and academia play a pivotal role in shaping the adoption of the Earthmate principles. On the positive side, scientists, researchers, and educators bring diverse perspectives and specialized knowledge, contributing to a comprehensive understanding of the principles emphasized by Earthmate. Environmental scientists delve into the intricate relationships within ecosystems, while psychologists explore the human psyche's response to ecological challenges. However, the potential drawback lies in the diversity within academia itself. Disparate views and conflicting interests among experts may lead to fragmented adoption of Earthmate, hindering the unified front required for its successful implementation. Striking a balance between individual expertise and a collective commitment to the Earthmate principles becomes crucial to mitigate these challenges.

Environmental Writers, Filmmakers and Documentarians

Environmental writers, filmmakers, and documentarians wield considerable influence in shaping public perceptions and fostering the adoption of the Earthmate principles. Positively, their creative works can effectively communicate the principles emphasized by Earthmate, fostering a deeper understanding of humanity's role in the larger ecological framework. Climate change documentarians, for instance, bring attention to the urgency of addressing environmental issues through compelling visuals and narratives. However, a potential drawback lies in the risk of oversimplification or sensationalism for the sake of engagement. Advocacy-focused filmmakers may inadvertently skew perspectives, focusing on extremes rather than presenting a nuanced view of the delicate balance Earthmate seeks to maintain. Striking a balance between raising awareness and maintaining the integrity of the principles becomes crucial for the positive impact of these mediums.

Celebrities, Influencers, Public Figures

Celebrities, influencers, and public figures play a significant role in influencing public perceptions and behaviors towards adopting the Earthmate principles. On the positive side, eco-friendly celebrities who genuinely align with the Earthmate principles can leverage their fame to raise awareness and garner support for environmental causes. Sustainable fashion influencers, eco-conscious bloggers, and environmental podcasters contribute by showcasing practical ways to integrate the Earthmate principles into daily life. However, a potential negative role emerges when celebrities engage in greenwashing, using environmental causes merely as a branding strategy without genuine commitment. This can dilute the authenticity of the Earthmate principles and lead to skepticism among the public. The challenge lies in distinguishing between authentic advocates and those engaging in superficial environmental gestures for personal gain.

Religious and Ethical Leaders

Religious and ethical leaders have the potential to play a pivotal role in adopting the Earthmate principles, aligning spiritual values with environmental stewardship. Green spiritual leaders, such as Pope Francis, have made significant strides in promoting ecological responsibility within their faith communities. Interfaith environmental advocates bring together diverse religious traditions, fostering a shared commitment to the Earthmate principles. The positive impact lies in the potential to influence millions of followers toward sustainable practices. However, challenges may arise when religious doctrines conflict with certain aspects of the Earthmate principles, leading to resistance or skepticism within some communities. Striking a balance between spiritual teachings and contemporary ecological concerns becomes crucial to maximize the positive influence of religious and ethical leaders.

Environmental Reporters and Journalists

Environmental reporters and journalists play a crucial role in shaping public awareness and understanding of the Earthmate principles. On the positive side, solutions-focused journalists contribute by highlighting practical applications of the Earthmate principles, inspiring communities and individuals to adopt more sustainable practices. Environmental photojournalists utilize powerful imagery to evoke empathy and connection, emphasizing the principles' core values of interconnectedness. However, a potential drawback is the ethical responsibility of journalists in their reporting. Striking a balance between accurate representation and avoiding sensationalism becomes crucial to ensure that the Earthmate principles is conveyed authentically. There's a risk that exaggerated or distorted narratives may lead to misinterpretations, hindering the philosophy's effectiveness in motivating positive action.

Influential Individuals in Sustainable Business

In the realm of sustainable business practices, influential individuals play a pivotal role in shaping the adoption of the Earthmate principles. On the positive side, green entrepreneurs prioritize the interconnectedness of business operations with ecological well-being, demonstrating how economic endeavors can coexist harmoniously with the planet. Sustainable tech innovators develop technologies that align with the Earthmate principles, emphasizing the interconnected nature of technological advancements and environmental responsibility. However, a potential drawback lies in the risk of greenwashing, where individuals may adopt sustainability as a mere marketing strategy without a genuine commitment to Earthmate values. Striking a balance between genuine environmental stewardship and profit motives becomes crucial to ensure that sustainable business practices align authentically with the philosophy's principles.

Environmental Activists and Advocates

Environmental activists and advocates serve as catalysts for change in adopting the Earthmate principles, contributing both positively and, at times, facing challenges. On the positive side, conservation campaigners actively engage in preserving biodiversity and ecosystems, aligning with the Earthmate principles of responsible stewardship. Affected community advocates highlight the interconnectedness of environmental issues with social justice concerns, bridging the gap between the Earthmate principles and human welfare. Policy advocates for environmental justice work towards systemic change, emphasizing the need for a balanced, interconnected approach. However, a potential drawback lies in the diversity of activism styles, with clashes sometimes occurring between different approaches. Balancing the urgency of direct action with the patience required for policy change becomes essential to ensure a cohesive and effective adoption of the Earthmate principles.

Environmental Justice and Climate Justice Advocates

Environmental justice and climate justice advocates play a vital role in championing the Earthmate principles, emphasizing the interconnectedness of social and environmental issues. Positively, activists addressing environmental racism highlight the interconnected nature of systemic discrimination and ecological degradation. Gender and environment advocates underscore the importance of considering the diverse ways environmental issues impact individuals based on their gender, further aligning with the Earthmate principles. Indigenous rights and land defenders, by protecting their ancestral lands, embody Earthmate's emphasis on responsible stewardship. However, a potential drawback exists in the challenge of reconciling the urgency of addressing environmental injustices with the need for systemic change. Striking a balance between immediate action and long-term, sustainable solutions becomes crucial to ensure that Earthmate values guide advocacy efforts effectively.

Culinary Experts and Food Influencers and Their Recipes

Culinary experts and food influencers wield significant influence in shaping dietary choices and culinary practices, presenting both positive and negative aspects in adopting the Earthmate principles. Positively, sustainable cooking channels offer innovative ways to align culinary practices with the Earthmate principles. These platforms emphasize the interconnectedness of food choices with environmental sustainability, promoting locally sourced, seasonal ingredients and minimizing food waste. However, a potential drawback exists when culinary experts and influencers focus on trendy diets or exotic ingredients that may contribute to overconsumption and environmental strain. Striking a balance between promoting Earthmate-inspired sustainable practices and engaging content that resonates with a broad audience becomes crucial for ensuring a positive impact.

Traditional Knowledge Holders

Traditional knowledge holders, including indigenous elders, traditional herbalists, and indigenous artisans, play a crucial role in adopting the Earthmate principles. Positively, indigenous elders and wisdom keepers often possess deep insights into the interconnectedness of nature and humanity, aligning with the Earthmate principles. Traditional herbalists contribute by promoting sustainable and holistic approaches to health, recognizing the intricate balance between human well-being and environmental health. Indigenous artisans, through sustainable practices, showcase the interconnected relationship between cultural expressions and the environment. However, a potential drawback arises when traditional knowledge is exploited or appropriated without proper understanding or respect. Striking a balance between sharing traditional knowledge for the greater good and protecting it from misuse becomes essential to ensure a positive alignment with the Earthmate principles.

Educational and Consultative Institutions

Schools and Environmental Education Programs

Schools and environmental education programs can contribute significantly to adopting the Earthmate principles. Positively, incorporating the Earthmate principles into the education system fosters an understanding of interconnectedness and responsibility toward the environment. Innovative curricula at various education levels, such as primary, secondary, and tertiary, provide a comprehensive approach to instill Earthmate values. Environmental education programs that integrate interdisciplinary learning across subjects offer a holistic understanding of the interconnected nature of ecological systems. However, a potential drawback lies in the variation in the implementation of environmental education, with some schools focusing more on theoretical knowledge than practical, hands-on experiences. Striking a balance between theoretical understanding and practical application becomes crucial to ensure that students not only comprehend the Earthmate principles but also internalize them through real-world experiences.

Colleges and Universities and Environmental Education Programs

Colleges and universities play a pivotal role in adopting the Earthmate principles and science, acting as crucibles for cultivating a generation attuned to the interconnectedness of existence. Positively, institutions with dedicated departments or faculties for environmental studies and sustainability contribute significantly to the Earthmate principles. These departments provide students with specialized knowledge and interdisciplinary perspectives, fostering a nuanced understanding of the delicate balance between individual autonomy and communal welfare. However, a potential drawback lies in the variation in the depth of environmental education across institutions, with some placing a stronger emphasis on traditional disciplines. Ensuring a uniform integration of the Earthmate principles across all disciplines becomes imperative to cultivate a comprehensive and holistic worldview.

Informal Online or Offline Environmental Education

Informal online or offline environmental education plays a vital role in disseminating the Earthmate principles to diverse audiences, fostering a global understanding of interconnectedness and responsible stewardship. Positively, these initiatives cater to specific age groups, including early childhood environmental programs, ensuring that environmental awareness is instilled from an early age. Platforms offering online courses and educational content for a global audience contribute to a democratized access to environmental knowledge. However, a potential drawback lies in the varying quality and reliability of information across different platforms, emphasizing the need for critical evaluation and fact-checking. Ensuring the accuracy of information disseminated through these channels becomes crucial to prevent the spread of misinformation.

Partnerships and Collaborations Targeting Environmental Education

Partnerships and collaborations are instrumental in advancing the Earthmate principles through environmental education. Positively, these collaborations foster a synergy between educational institutions and industry partners, leading to applied research with real-world impact. The integration of traditional ecological knowledge into environmental education programs exemplifies the positive role of partnerships in embracing diverse perspectives. However, potential challenges include the risk of industry influence affecting educational content. Striking a balance between academic integrity and practical relevance becomes crucial to ensure that collaborative efforts uphold the Earthmate principles, avoiding any compromise on scientific accuracy and environmental ethics.

Environmental Consultants and Advisory Services

Environmental consultants play a pivotal role in advancing the Earthmate principles by providing tailored solutions to businesses, government agencies, and NGOs. Positively, their expertise aids organizations in adopting sustainable practices and minimizing environmental impact. For instance, consultants specializing in sustainability contribute to the integration of circular economy principles, promoting responsible resource consumption and regenerative production. However, potential challenges include instances where consultants may prioritize short-term economic interests over long-term environmental benefits. Striking a balance between economic prosperity and ecological responsibility becomes crucial, ensuring that consultancy practices align with the Earthmate principles, fostering holistic and sustainable solutions.

Think Tanks and Policy Institutes

Think tanks and policy institutes play a crucial role in adopting the Earthmate principles by contributing to the development of policies rooted in the philosophy and science of environmental sustainability. Positively, these institutions engage in in-depth research, providing valuable insights into climate policy, conservation strategies, and sustainable development. They act as catalysts for change by influencing public discourse and shaping environmental policies at various levels. However, potential challenges include the risk of think tanks being influenced by political or economic interests, which might compromise the purity of the Earthmate principles. Maintaining transparency, avoiding conflicts of interest, and prioritizing the long-term well-being of the planet are essential for ensuring that think tanks effectively contribute to the Earthmate principles.

R&D Institutions

R&D institutions play a pivotal role in adopting the Earthmate principles by driving innovation and technological advancements aimed at addressing environmental challenges. Positively, these institutions contribute to the development of cutting-edge technologies for environmental monitoring, climate adaptation, and resource conservation. They foster a culture of innovation, encouraging scientists and researchers to explore novel solutions aligned with the Earthmate principles and science. However, potential challenges include the risk of prioritizing short-term gains over long-term ecological sustainability. R&D institutions must navigate economic pressures and ensure that technological advancements align with Earthmate's commitment to responsible stewardship of the planet.

Media and Culture

Media and Communication

Media and communication can serve as powerful tools in adopting the Earthmate principles, acting as both positive influencers and potential challenges. Positively, media outlets, especially those focusing on environmental issues, play a crucial role in shaping public awareness and policy discourse. They disseminate information, fostering global cooperation and influencing responsible behaviors. However, challenges arise when mainstream media outlets prioritize sensationalism or fail to provide comprehensive coverage of environmental topics. Striking a balance between attracting audiences and promoting accurate, unbiased information is essential for media outlets aligning with the Earthmate principles.

Social Media and Online Activism

Social media and online activism serve as powerful tools for fostering environmental awareness and mobilizing global citizens. On the positive side, these platforms enable individuals worldwide to unite in advocating for environmental causes, creating a shared sense of responsibility. The Earthmate principles, emphasizing unity despite plurality, find expression in the diverse yet interconnected voices amplified through social media. Online activism allows for the rapid dissemination of information, facilitating real-time responses to environmental crises. However, the negative aspects include the potential for misinformation, "clicktivism" or superficial engagement, and the risk of oversimplifying complex environmental issues. Maintaining a balance between impactful activism and avoiding the pitfalls of online discourse remains a challenge.

Religious and Ethical Perspectives

Religious and ethical perspectives play a pivotal role in shaping attitudes toward the environment. On the positive side, many religious teachings emphasize the sacredness of nature and humanity's role as stewards of the Earth. For example, the concept of stewardship in Christianity, the interconnectedness in Buddhism, or the reverence for nature in indigenous beliefs align with the Earthmate principles. These perspectives often contribute to a sense of environmental responsibility and can foster a collective commitment to sustainable practices. However, challenges exist, such as rigid interpretations that may hinder adaptation to evolving environmental challenges or conflicts arising from divergent ethical beliefs. Striking a balance between adhering to traditional principles and adapting to modern environmental imperatives remains a challenge.

Art and Literature

Art and literature serve as powerful mediums in promoting the Earthmate principles. On the positive side, environmental art transcends traditional communication methods, conveying complex issues through visual and sensory experiences. This art often fosters a deep emotional connection with nature, inspiring sustainable practices and fostering a sense of responsibility. The fashion and design industry plays a pivotal role by increasingly embracing sustainable and ethical practices, such as using eco-friendly materials and ethical supply chains. However, challenges exist, including potential greenwashing in the fashion industry, where companies may present a false image of sustainability. Balancing artistic freedom with responsible environmental messaging is a delicate task, as misrepresentation or oversimplification can dilute the urgency of environmental issues.

Profit-Oriented Entities

Financial Institutions

Financial institutions play a pivotal role in either hindering or advancing the Earthmate principles. As highlighted in a UNEP report (2023), finance flows to NbS, amounting to US$200 billion, are massively outweighed by nearly US$7 trillion in flows with direct negative impacts on nature, emphasizing the significant role of financial activities in environmental degradation. On the positive side, sustainable finance practices—such as green banking and ethical investment portfolios—demonstrate a growing commitment to aligning financial activities with environmental responsibility. Institutions adopting these practices contribute to the transition toward a more sustainable economy. However, challenges remain, particularly the reluctance of traditional financial entities to fully embrace these principles. When profitability is prioritized over sustainability, the result can be a negative impact on Earth's interconnected systems.

Industrial Productors

Industrial producers play a pivotal role in either contributing positively or negatively to the Earthmate principles. On the positive side, certain industries actively engage in sustainable practices, contributing to environmental conservation and climate change mitigation. Recognizing and distinguishing between industries based on their environmental impact is crucial, especially for sectors actively working towards implementing eco-friendly production processes and circular economy practices. However, challenges persist, with some industries lagging in adopting sustainable practices, leading to negative consequences for the interconnected systems of Earth.

Mining Industry

Certain segments of the mining industry have actively embraced the Earthmate principles by implementing sustainable mining practices. These practices often include comprehensive land reclamation initiatives, where mining companies rehabilitate and restore mined areas, ensuring that ecosystems are revitalized. Additionally, community engagement initiatives foster positive relationships between mining companies and local communities, emphasizing the interconnectedness between economic activities and social well-being. The responsible sourcing of materials is another positive facet, aiming to minimize environmental and social impacts throughout the supply chain. By adhering to these Earthmate principles, some mining companies contribute positively to the delicate balance between economic prosperity and environmental protection. However, despite these efforts, the inherent nature of large-scale mining operations can still lead to significant environmental degradation, such as deforestation, water contamination, and biodiversity loss, underscoring the challenge of fully aligning such activities with sustainability goals.

Agriculture, Agroforestry, Permaculture, Sustainable Agriculture and Forestry Initiatives

The adoption of agroecological practices and the promotion of biodiversity-friendly farming are crucial in mitigating the negative impacts of conventional agriculture on ecosystems. Techniques such as organic farming, agroforestry, and permaculture contribute significantly to the preservation of biodiversity, soil health, and water conservation. As noted by Smil (1994), there should be a shift beyond the self-serving altruism of wealthy nations towards the increased dissemination of efficient farming techniques and substantial food aid during critical emergencies to developing countries. This shift, combined with a move towards sustainable diets domestically, is essential to ensure a sufficient global food supply and significantly reduce the "affluence gap" and associated fertility disparities between developed and developing regions. Sustainable agriculture practices, rooted in the Earthmate principles, emphasize harmony between human activities and nature, fostering a resilient and regenerative approach to food production. Furthermore, sustainable forestry initiatives enhance these efforts by promoting responsible land management and forest conservation. By recognizing and supporting companies that prioritize these practices, a transition towards environmentally friendly and socially responsible agricultural methods can be effectively promoted. However, a potential challenge in the adoption of these sustainable practices is the initial costs and labor-intensive nature of transitioning, which may deter small-scale farmers. This situation could lead to economic challenges and reduced agricultural productivity in the short term, ultimately hindering the broader goal of sustainable food systems.

Technological Innovators and Environmental Technology Startups

Technological innovators and environmental technology startups play a pivotal role in advancing the Earthmate principles. Their positive impact is evident in the development of sustainable solutions that address environmental challenges. For instance, companies focused on renewable energy innovations contribute significantly to reducing dependence on fossil fuels, mitigating climate change, and promoting cleaner energy sources. The advent of smart city technologies enhances urban sustainability by optimizing resource use, improving energy efficiency, and creating resilient urban ecosystems. Sustainable mobility solutions, including electric vehicles and intelligent transportation systems, exemplify the positive role of technology in reducing carbon emissions and fostering eco-friendly transportation alternatives. However, a potential drawback lies in the environmental costs of producing and disposing of high-tech products, which may involve rare earth mining, electronic waste, and a significant carbon footprint during manufacturing, potentially undermining the overall sustainability goals of such innovations.

Tourism Industry, Sustainable Tourism and Ecotourism

The tourism industry plays a significant role in both harming and advancing environmental sustainability. Sustainable tourism emphasizes responsible travel that benefits local communities and minimizes environmental damage, setting itself apart from the negative impacts of mass tourism. By prioritizing environmental conservation and cultural preservation, sustainable tourism aligns with the Earthmate principles, recognizing the interconnectedness of ecosystems and local communities. Ecotourism, in particular, highlights practices that protect biodiversity while fostering education and awareness among travelers. Tourism operators committed to reducing their environmental footprint can promote conservation efforts and improve the well-being of destinations. Collaborations with indigenous and local communities further enhance the positive impact of sustainable tourism by ensuring these communities are active participants in conservation initiatives. However, a potential drawback of the tourism industry, even in the context of sustainable tourism and ecotourism, is the risk of over-commercialization. Increased tourist traffic in delicate ecosystems can strain local resources and lead to environmental degradation, undermining the very principles that sustainable and eco-friendly tourism seek to promote.

Supply Chains and Consumer Behavior

Supply chains play a pivotal role in adopting the Earthmate principles, particularly in promoting sustainability and ethical production. Supply chain transparency is a key driver, allowing consumers and businesses to trace the origins of products, ensuring they meet ethical and environmental standards. Companies adopting circular supply chain models contribute significantly to Earthmate goals by minimizing waste and promoting resource efficiency. Emphasizing sustainable sourcing and fair labor practices within supply chains aligns with Earthmate's call for responsible stewardship. When companies prioritize these principles, supply chains become a powerful mechanism for driving positive environmental and social change. However, the complexity of global supply chains can lead to challenges in fully ensuring ethical standards are met at every stage, and in some cases, the costs associated with implementing sustainable practices may be passed on to consumers, potentially limiting access to eco-friendly products.

Green Building and Construction

Green building and construction practices play a pivotal role in adopting the Earthmate principles by aligning with the philosophy's emphasis on responsible stewardship of the planet. Companies in the construction sector that prioritize sustainable building practices, energy efficiency, and the use of eco-friendly materials contribute to the interconnected goals of the Earthmate principles and science. Such practices, including the incorporation of energy-efficient designs and the use of renewable energy sources, demonstrate a commitment to reducing the environmental impact of construction projects. The promotion of circular economy principles within the construction sector, focusing on responsible resource consumption and a regenerative approach to production, further aligns with Earthmate's vision of a sustainable and ecologically balanced future. However, the higher upfront costs associated with sustainable construction methods and materials may deter some developers, particularly in regions with limited financial resources, slowing the widespread adoption of eco-friendly building practices.

Packaging and Sustainable Packaging Companies

Sustainable packaging companies play a crucial role in adopting the Earthmate principles by aligning with the philosophy's emphasis on responsible stewardship of the planet. Companies that prioritize sustainable packaging solutions contribute to Earthmate's vision of a delicate balance between economic prosperity, social well-being, and environmental protection. By offering alternatives to single-use plastics and materials with lower environmental impact, these companies address the interconnected challenges of resource consumption and waste generation. The adoption of circular economy principles, where materials are reused or recycled, further supports Earthmate's call for a regenerative approach to production. As these companies actively engage in preventing environmental harm and conserving resources, they exemplify the interconnectedness of responsible business practices with the broader goals of the Earthmate principles. However, the higher costs of producing sustainable packaging compared to conventional options may deter widespread adoption, especially among smaller businesses, potentially slowing progress toward large-scale environmental improvements.

Waste Management Industry

The waste management industry plays a pivotal role in adopting the Earthmate principles by addressing environmental challenges through sustainable practices. Companies within this sector that prioritize recycling, waste reduction, and resource recovery contribute significantly to Earthmate's vision of responsible stewardship. According to analyses of Maalouf and Mavropoulos (2023), total global waste arisings amounted to approximately 20 billion tonnes in 2017, which corresponds to 2.63 tonnes of waste per capita per year. Under a business-as-usual scenario, global waste is expected to grow to 46 billion tonnes by 2050, highlighting the immense scale of waste management challenges. By diverting waste from landfills and implementing recycling programs, these companies align with Earthmate's call for a delicate balance between economic prosperity and environmental protection. Innovations in waste-to-energy technologies further showcase the industry's commitment to a regenerative approach, transforming waste into a valuable resource. Waste management companies actively engaged in reducing single-use plastics exemplify the interconnected goals of the Earthmate principles, promoting circular economy principles and responsible resource consumption. However, despite these efforts, the industry faces challenges in fully transitioning to sustainable practices, including the high costs and technical limitations of advanced recycling and waste-to-energy technologies, which may hinder large-scale adoption and the industry's overall impact on reducing waste generation.

Renewable Energy Industry

The renewable energy industry plays a vital role in aligning with the Earthmate principles by contributing to responsible stewardship of the planet. Companies within this sector are pivotal in the transition toward sustainable energy practices, emphasizing the interconnectedness between human activities and the environment. Providers that actively engage local communities in the development and benefits of renewable energy projects embody Earthmate's emphasis on collective well-being. By ensuring a fair distribution of positive impacts, these companies foster unity despite the plurality of interests, acknowledging the importance of balancing individual rights with communal welfare. The renewable energy industry, through its commitment to cleaner energy production, exemplifies Earthmate's call for a self-regulating, interconnected whole, where economic prosperity is harmonized with environmental responsibility. However, the expansion of renewable energy infrastructure, such as wind and solar farms, can sometimes result in unintended environmental and social consequences, including habitat disruption, land-use conflicts, and challenges to biodiversity conservation, potentially contradicting the broader environmental goals they aim to support.

PPPs

PPPs serve as a dynamic force in adopting the Earthmate principles by fostering collaboration between government entities and private-sector organizations. These partnerships contribute positively to Earthmate's call for collective action, emphasizing the interconnectedness of efforts to address environmental challenges. Collaborations between profit-oriented entities and governmental or non-profit organizations represent a harmonious blend of individual autonomy and communal welfare. PPPs play a pivotal role in financing and implementing conservation projects, promoting responsible stewardship of the planet. These partnerships often lead to the development of innovative solutions that align with Earthmate's commitment to sustainable development principles. By bridging gaps and leveraging the strengths of both sectors, PPPs exemplify the philosophy's vision of unity despite the diversity of interests. However, the involvement of private-sector entities in PPPs can sometimes result in prioritizing profitability over long-term sustainability, potentially leading to compromises in environmental standards or the marginalization of local communities, thus undermining the broader goals of equitable and responsible stewardship.

Non-profit Entities

Environmental Justice and Climate Justice Associations

Environmental Justice and Climate Justice Associations play a crucial role in embodying the Earthmate principles. These organizations become champions for the interconnectedness between environmental well-being, social justice, and responsible stewardship, aligning with Earthmate's emphasis on unity despite diversity. By empowering communities to actively participate in environmental decision-making processes, these associations promote a nuanced understanding of the interplay between individual rights and communal welfare. Through their work, they navigate the philosophical terrain outlined by Earthmate, recognizing the need for a balanced and collective approach. These associations serve as catalysts for achieving a self-regulating, interconnected whole, where marginalized communities and the broader society collaboratively engage in sustainable practices. However, a potential negative effect arises when these associations face opposition from powerful entities that may attempt to suppress or undermine their efforts, hindering progress and perpetuating inequalities, especially when environmental justice goals conflict with economic interests.

Wildlife Conservation Associations

Wildlife Conservation Associations play a vital role in embodying the Earthmate principles. By focusing on biodiversity conservation, habitat preservation, and species recovery, these associations align with Earthmate's emphasis on the interconnectedness of biological and ecological systems. Successful practices and initiatives include comprehensive programs that extend beyond species protection to ecosystem restoration. However, despite their positive impact, certain conservation efforts may unintentionally disrupt local communities' livelihoods or lead to imbalanced resource allocation, potentially causing conflicts. These associations contribute to Earthmate's vision by recognizing the delicate balance required for the coexistence of diverse species and ecosystems. Initiatives that integrate local communities into wildlife conservation strategies are particularly impactful, showcasing Earthmate's acknowledgment of the interconnected nature of sociological and ecological realms.

Climate Change Mitigation and Adaptation Associations

Organizations dedicated to climate change mitigation and adaptation play a crucial role in embodying the Earthmate principles. By addressing the interconnected challenges of climate change through scientific, technological, and societal approaches, these associations align with Earthmate's holistic vision. The distinction between organizations focused on mitigating the causes of climate change and those specializing in adaptation strategies recognizes the multifaceted nature of environmental crises. However, certain mitigation strategies, if not carefully managed, could lead to economic imbalances or unintended social consequences, particularly in vulnerable communities. The Earthmate principles encourage the delicate balance between mitigating impacts and adapting to changing conditions, emphasizing the interconnectedness of environmental, social, and economic systems. Successful practices showcase the integration of innovative approaches to climate resilience and adaptation, contributing to Earthmate's call for responsible stewardship of the planet.

Ocean Protection Associations

Ocean Protection Associations play a pivotal role in aligning with the Earthmate principles by addressing the positive and negative aspects of their impact on marine ecosystems. On the positive side, these associations emphasize the critical importance of ocean conservation, highlighting issues such as overfishing, plastic pollution, and coral reef degradation. By actively engaging in initiatives to combat these challenges, they embody Earthmate's call for responsible stewardship of the planet's resources. However, certain conservation efforts, if not carefully implemented, can lead to unintended consequences, such as disrupting local economies or displacing communities reliant on marine resources. Collaborations with local communities for sustainable marine resource management further exemplify their commitment to the Earthmate principles by fostering inclusivity and shared responsibility.

Community-Based Conservation Initiatives

Community-Based Conservation Initiatives play a pivotal role in embodying the Earthmate principles through their positive contributions and potential challenges. On the positive side, these initiatives align with Earthmate's call for responsible stewardship of the planet by recognizing the interconnectedness of nature and human societies. Many grassroots organizations work closely with local communities, fostering sustainable resource management and conservation practices. Through community engagement and empowerment, these initiatives ensure that conservation efforts are not only effective but also consider the well-being of the communities directly impacted. However, if not managed inclusively, such efforts can inadvertently marginalize certain groups or disrupt traditional livelihoods, creating tensions within the communities they aim to support.

Green Philanthropy and Impact Investing

Green Philanthropy and Impact Investing play a pivotal role in adopting the Earthmate principles, offering both positive contributions and potential challenges. On the positive side, these financial approaches align with Earthmate's call for a delicate balance between economic prosperity, social well-being, and environmental protection. Organizations engaged in green philanthropy and impact investing contribute significantly to funding innovative environmental initiatives. These initiatives aim to address the multifaceted challenges posed by environmental crises, such as climate change and biodiversity loss, by providing crucial financial support to projects that align with Earthmate's holistic approach to sustainability. However, a potential negative effect is that such investments may sometimes prioritize profit-driven outcomes over long-term environmental benefits, risking the dilution of Earthmate's core principles in favor of short-term financial gains.

Legal Advocacy and Environmental Law NGOs

Legal advocacy and Environmental Law NGOs play a crucial role in adopting the Earthmate principles, bringing both positive and negative aspects to the forefront. On the positive side, these organizations contribute significantly to Earthmate's call for responsible stewardship of the planet. Through legal actions and advocacy, they act as guardians of environmental protection, holding individuals and entities accountable for violations against nature. By engaging in litigation against environmental violations, these organizations align with the Earthmate principles, emphasizing the interconnectedness and interdependence of human activities with the well-being of the planet. However, a potential negative effect is that excessive reliance on legal approaches can sometimes lead to prolonged litigation processes, diverting resources and attention away from proactive, collaborative environmental solutions and hindering timely action.

Crisis Response and Disaster Relief NGOs

Crisis Response and Disaster Relief NGOs play a pivotal role in adopting the Earthmate principles, presenting both positive and negative dimensions. On the positive side, these organizations embody Earthmate's call for responsible stewardship by providing swift and effective assistance during natural disasters and climate-related emergencies. Through their rapid response efforts, they exemplify the interconnectedness of humanity and the environment, showcasing a collective commitment to addressing the immediate needs of affected communities and ecosystems. Rapid response to natural disasters, such as hurricanes, floods, and wildfires, aligns with the Earthmate principles of collective well-being and interconnectedness. By addressing the immediate needs of affected communities and ecosystems, these NGOs contribute to the philosophy's emphasis on unity despite diversity, demonstrating a shared responsibility for the planet's well-being. However, a potential negative effect is that in focusing on short-term relief, these organizations may inadvertently overlook long-term, sustainable solutions to disaster prevention and resilience, potentially reinforcing reactive approaches rather than proactive environmental strategies.

Governmental Entities

Governmental entities play a profound and decisive role in adopting and enforcing the Earthmate principles, shaping the future of sustainable development and environmental stewardship. These entities have the capacity to drive large-scale change by setting policies, regulating industries, and allocating resources toward initiatives that align with Earthmate's vision of balancing economic, social, and environmental well-being. If governance were structured around the Four Freedoms—freedom of speech, freedom of worship, freedom from want, and freedom from fear—not only would social and political justice be deeply embedded in the system, but it would also provide a foundation for a more equitable and inclusive society. This form of governance would promote human dignity, equality, and sustainability.

Furthermore, if future governance models are enhanced by artificial intelligence systems built on the principles of transparency, fairness, and impartiality—free from bias, corruption, discrimination, and injustice—the potential for addressing fundamental issues would be exponentially increased. AI-driven governance could optimize resource distribution, ensure accountability in decision-making processes, and provide real-time solutions to pressing environmental crises. Such a system could proactively prevent problems rather than merely react to them, fostering a harmonious balance between human progress and the planet's ecological integrity.

In this future scenario, AI governance would enhance the precision and effectiveness of policy-making, providing data-driven insights into climate adaptation, environmental protection, and economic sustainability. Issues such as poverty, inequality, environmental degradation, and inefficient resource management would be mitigated through intelligent systems capable of unbiased decision-making. By removing human flaws from governance, including personal biases, mismanagement, or susceptibility to special interests, AI could help create a system where decisions are made for the collective good, addressing root causes rather than symptoms, and leading humanity toward a more just, sustainable, and prosperous future.

Legislative Bodies

Legislative bodies play a crucial role in adopting the Earthmate principles, presenting both positive and negative aspects. On the positive side, legislative bodies serve as key institutions for developing, reviewing, and passing environmental policies and regulations. They play a vital role in addressing the interconnected challenges of environmental degradation and climate change by formulating laws that guide sustainable development. Highlighting instances of bipartisan or cross-party collaboration within legislative bodies showcases the potential for effective and sustainable environmental policies, emphasizing the need for consensus in addressing complex issues. A potential negative effect of legislative bodies in adopting the Earthmate principles is that environmental policies may be delayed or weakened due to political gridlock, lobbying from powerful interest groups, or lack of consensus among lawmakers. These challenges can hinder timely and effective responses to urgent environmental crises, such as climate change and biodiversity loss. In some cases, political polarization can lead to the dilution of key sustainability measures, limiting their impact and undermining long-term environmental goals.

Legal and Regulatory Frameworks

Legal and regulatory frameworks play a pivotal role in shaping and implementing the Earthmate principles. On the positive side, stringent and effectively enforced legal frameworks can act as powerful tools in addressing environmental challenges. They provide a structured approach to ensuring responsible resource consumption, waste management, and biodiversity conservation. The adaptability of legal frameworks to emerging environmental issues, such as technological advancements or shifts in climate patterns, showcases their potential to evolve with the changing landscape. Recognizing entities that have developed comprehensive legal frameworks addressing various environmental aspects highlights the importance of a holistic approach within regulatory systems. A potential negative effect of legal and regulatory frameworks in adopting the Earthmate principles is that overly complex or rigid regulations may stifle innovation or create burdensome compliance costs for businesses and governments. In some cases, these frameworks can be slow to adapt to rapidly evolving environmental issues or new technologies, resulting in outdated laws that fail to address emerging environmental challenges effectively. Additionally, weak enforcement or loopholes in the legal system may allow non-compliance, undermining the overall effectiveness of environmental protection efforts.

Law Enforcement Entities

Law enforcement entities play a critical role in adopting the Earthmate principles, contributing both positively and negatively to environmental protection. On the positive side, specialized units within law enforcement agencies dedicated to investigating and combating environmental crimes, such as illegal logging, wildlife trafficking, and pollution, represent a proactive approach to addressing ecological threats. Collaborative efforts between law enforcement and local communities amplify the positive impact by fostering shared responsibility and engagement in environmental protection. Additionally, law enforcement efforts to combat environmental crimes demonstrate a commitment to upholding environmental regulations and protecting biodiversity. A potential negative effect of law enforcement entities in adopting the Earthmate principles is that their approach to environmental protection can sometimes lead to conflicts with local communities, particularly if enforcement actions are perceived as heavy-handed or unjust. Such tensions can result in resistance from those who feel that their livelihoods are threatened by strict regulations or enforcement practices. Moreover, if law enforcement agencies lack adequate training or resources to effectively address environmental issues, it may lead to inconsistent enforcement and a lack of accountability, undermining trust in these entities and their role in protecting the environment.

Judiciary

The judiciary plays a crucial positive role in adopting the Earthmate principles by establishing specialized courts or tribunals for handling environmental cases. Jurisdictions that recognize the unique challenges of environmental issues often set up dedicated legal bodies with expertise in environmental matters. For instance, countries like India have established specialized environmental courts, such as the National Green Tribunal, to expedite the resolution of environmental disputes. These specialized courts ensure a more nuanced understanding of scientific and ecological complexities, fostering effective legal decisions aligned with the Earthmate principles. A potential negative effect of the judiciary in adopting the Earthmate principles is that specialized environmental courts may face challenges related to accessibility and resource allocation, which can hinder their effectiveness. For instance, if these courts are underfunded or lack sufficient personnel, they may struggle to manage caseloads efficiently, leading to delays in justice and resolutions of environmental disputes. Furthermore, disparities in access to these courts can arise, potentially favoring well-resourced entities over marginalized communities, which undermines the equitable application of environmental laws. This inequity can perpetuate environmental injustices, contradicting the foundational principles of Earthmate, which emphasize holistic and inclusive approaches to sustainability.

Governmental Organizations for Wildlife Conservation

Governmental organizations dedicated to wildlife conservation play a vital role in adopting the Earthmate principles by managing and conserving wildlife in protected areas. National parks and reserves, overseen by these entities, serve as crucial sanctuaries for diverse ecosystems. For example, the Yellowstone National Park in the United States is managed by the National Park Service, emphasizing the preservation of biodiversity and natural processes. By safeguarding these areas, governmental organizations contribute to the Earthmate principles, recognizing the interconnectedness of all living beings and the need for responsible stewardship. A potential negative effect of governmental organizations dedicated to wildlife conservation is that their management practices can sometimes lead to conflicts with local communities, particularly if conservation efforts restrict traditional land use or access to resources. These restrictions may create resentment among locals who rely on these areas for their livelihoods, leading to tensions that can undermine collaborative conservation efforts. Additionally, if these organizations prioritize certain species over others or focus solely on charismatic megafauna, they may inadvertently neglect the needs of less visible but equally important species, which can disrupt ecosystem balance and counteract the holistic principles of Earthmate.

Government Water Resource Management Entities

Governmental water resource management entities play a crucial role in adopting the Earthmate principles through the encouragement of integrated water resource management plans. By developing and implementing such plans, these entities prioritize the needs of both ecosystems and communities. For instance, the Murray-Darling Basin Authority in Australia has adopted IWRM principles to balance water allocations between agriculture, urban areas, and the environment. This approach reflects the Earthmate principles by recognizing the interconnectedness of water systems, human communities, and the broader ecosystem. A potential negative effect of governmental water resource management entities is that their integrated water resource management plans may inadvertently prioritize the interests of powerful agricultural or industrial stakeholders over the needs of local communities and ecosystems. This can lead to inequities in water allocation, where marginalized groups face water scarcity while larger entities benefit disproportionately. Additionally, if these plans are not based on comprehensive data or fail to incorporate indigenous knowledge and community input, they may overlook critical ecological considerations and create long-term sustainability challenges. Such misalignments can ultimately undermine the holistic approach that the Earthmate principles advocate, threatening both social equity and environmental integrity.

Governmental Institutions for Meteorology, Climate Science, and Weather Forecasting

Governmental institutions dedicated to meteorology, climate science, and weather forecasting play a pivotal role in adopting the Earthmate principles through their contributions to climate science and weather forecasting. These institutions actively engage in climate modeling, impact assessments, and the development of adaptation strategies. For example, NASA and NOAA in the United States conduct extensive research to understand the Earth's climate system. Their efforts align with the Earthmate principles by recognizing the interconnectedness of environmental factors and advocating for responsible stewardship of the planet. A potential negative effect of governmental institutions dedicated to meteorology, climate science, and weather forecasting is that, despite their crucial role in climate modeling and impact assessments, their predictions may sometimes be hampered by uncertainties or incomplete data. This can lead to inaccurate forecasts, which might result in inadequate preparation for extreme weather events or misguided climate adaptation strategies. Additionally, these institutions can face challenges in effectively communicating scientific findings to the public or policymakers, leading to skepticism, misunderstanding, or delayed action, which undermines the proactive environmental stewardship that the Earthmate principles advocate.

Environmental Government Agencies

Environmental government agencies play a crucial role in adopting the Earthmate principles. Positively, these agencies serve as custodians of environmental well-being by formulating and implementing policies rooted in the Earthmate principles. For instance, EPA in the United States emphasizes the interconnectedness of environmental systems and human activities in its policies, aligning with Earthmate's call for holistic approaches. However, challenges arise in balancing environmental protection with economic interests, often leading to compromises in regulations. Agencies may face political pressures, impacting their autonomy and effectiveness in safeguarding the environment. The delicate balance between individual rights and communal welfare, as advocated by the Earthmate principles, becomes challenging amid conflicting interests and limited resources. A potential negative effect of environmental government agencies is that they may become subject to bureaucratic inefficiencies and slow decision-making processes, which can delay the implementation of critical environmental policies. Additionally, these agencies may struggle with inadequate funding or staffing, limiting their ability to enforce regulations effectively. In some cases, political influence or lobbying from industries with economic interests can lead to weakened environmental protections or the rollback of regulations, undermining their mission to safeguard natural resources. These challenges can compromise the holistic approach advocated by the Earthmate principles, which emphasizes the need for strong, impartial governance to protect both ecosystems and communities.

Urban Planning, Green Infrastructure and Sustainable Development

Urban Planning, Green Infrastructure, and Sustainable Development play a pivotal role in adopting the Earthmate principles, shaping the physical and environmental aspects of human habitats. Positively, urban planning initiatives that incorporate green infrastructure contribute to the promotion of sustainability in city development. Governmental entities actively involved in such endeavors prioritize creating environmentally friendly and resilient urban spaces. For example, cities like Copenhagen and Singapore have embraced sustainable urban development practices by incorporating green roofs, vertical gardens, and efficient waste management systems. These initiatives align with the Earthmate principles, emphasizing responsible stewardship of the planet by integrating ecological considerations into the urban landscape. A potential negative effect of urban planning initiatives that incorporate green infrastructure is that such projects can sometimes lead to "green gentrification," where the development of environmentally friendly spaces increases property values and displaces lower-income residents. Additionally, implementing green infrastructure may require significant financial investment, which can strain public budgets or lead to inequitable distribution of resources, with wealthier areas benefiting more than poorer neighborhoods. This disparity can undermine the Earthmate principles of inclusivity and equity, as the social benefits of sustainable development are not always equally shared across all communities.

Health and Environment

Healthcare Industry

The healthcare industry plays a pivotal role in both positive and negative aspects concerning the adoption of the Earthmate principles. On the positive side, there is a growing recognition of the healthcare industry's environmental impact, prompting initiatives to assess and mitigate its footprint. Encouraging the healthcare sector to address waste generation, energy consumption, and the use of harmful chemicals aligns with the Earthmate principles of responsible stewardship. Healthcare facilities implementing sustainable practices, such as energy-efficient infrastructure and eco-friendly purchasing policies, exemplify a positive role in incorporating Earthmate values into their operations. However, challenges persist, including the negative impact of healthcare waste and the industry's carbon footprint. Proper waste disposal and reducing chemical exposures within healthcare settings are vital for mitigating these environmental hazards, requiring ongoing efforts for improvement.

Medical Infrastructure and Health Systems

Medical infrastructure and health systems play a dual role in the context of the Earthmate principles, contributing both positively and negatively to the interconnected web of environmental and human well-being. On the positive side, addressing health disparities in communities disproportionately affected by environmental hazards aligns with the Earthmate principles of environmental justice. Advocating for health policies that recognize and mitigate the disproportionate health impacts on marginalized and vulnerable communities demonstrates a commitment to the interconnected nature of human and environmental health. Additionally, incorporating climate resilience strategies within health systems to mitigate the health impacts of climate change exemplifies a positive step toward responsible stewardship and aligns with Earthmate's call for a delicate balance between economic prosperity, social well-being, and environmental protection. A potential negative effect of medical infrastructure and health systems is their significant environmental footprint, particularly through the generation of medical waste and high energy consumption. Hospitals and healthcare facilities often produce large quantities of hazardous waste, which can contribute to environmental degradation if not managed properly. Furthermore, the extensive use of single-use plastics and other non-recyclable materials within the healthcare industry exacerbates pollution, counteracting the Earthmate principles of sustainability and environmental stewardship. Addressing these environmental impacts while maintaining high-quality healthcare services presents a significant challenge, as balancing human health with environmental health can sometimes lead to conflicting priorities.

Environmental Psychology

Environmental psychology plays a pivotal role in shaping human interactions with the environment, with both positive and negative aspects influencing the adoption of the Earthmate principles. On the positive side, research and initiatives exploring the positive effects of exposure to nature on mental health and well-being contribute to the holistic vision of Earthmate. The therapeutic benefits of nature on mental health, supported by studies and initiatives, align with Earthmate's emphasis on interconnectedness and responsible stewardship. Encouraging collaboration between environmental psychologists and urban planners to design spaces that support mental health exemplifies a positive role in integrating psychological insights into the planning of sustainable and human-centric environments. A potential negative effect of environmental psychology is that while it highlights the benefits of nature on mental health, it may inadvertently create unequal access to these benefits. In urban settings, marginalized communities often have less access to green spaces, leading to disparities in the mental health benefits of nature. This unequal distribution of environmental resources undermines the Earthmate principles of equity and interconnectedness, as those in underserved areas may not experience the same psychological benefits from nature exposure. Additionally, an overemphasis on individual behavioral changes in response to environmental psychology might shift focus away from larger systemic issues that require collective action and policy reform.

Community-Based Health Initiatives

Community-based health initiatives play a vital role in embodying the Earthmate principles, showcasing both positive and negative aspects. On the positive side, these initiatives often empower local communities to address environmental and public health concerns, aligning with Earthmate's call for responsible stewardship and collective well-being. By fostering a sense of collective responsibility, these initiatives encourage communities to actively participate in the protection and improvement of their local environments. Successful projects have demonstrated the potential for transformative change when communities are engaged in both monitoring and improving local environmental conditions. A potential negative effect of community-based health initiatives is that they may lack the necessary resources or expertise to address complex environmental health challenges effectively. While grassroots involvement is crucial, such initiatives can sometimes struggle with inconsistent funding, limited technical knowledge, or fragmented coordination, which can hinder long-term sustainability. Additionally, over-reliance on local communities to solve environmental health issues might divert attention from larger systemic and governmental responsibilities, leading to gaps in comprehensive policy interventions that are essential for addressing the root causes of environmental health problems.

Global Health and Environmental International Organizations

Global health and environmental international organizations play a pivotal role in shaping the adoption of the Earthmate principles, with both positive and negative aspects. On the positive side, these organizations often serve as powerful advocates for sustainable practices, aligning with Earthmate's call for responsible stewardship and a holistic approach to global challenges. They contribute to global cooperation, bridging understanding gaps and fostering collaboration across nations. Many of them, such as the HEAL, showcase successful international initiatives that address the interconnected nature of health and the environment. HEAL, for instance, focuses on advocating for policies that promote a healthy environment, recognizing the interdependence of human health and the planet. A potential negative effect of global health and environmental international organizations is that their broad focus on global cooperation can sometimes lead to slow, bureaucratic decision-making, which may delay urgent actions. Additionally, these organizations often face challenges in reconciling the diverse priorities and interests of member nations, which can result in diluted or compromised policies that may not fully address pressing environmental and health concerns. Moreover, the reliance on voluntary commitments from countries can limit the effectiveness of international initiatives, as some governments may prioritize economic growth over environmental sustainability, hindering progress toward the Earthmate principles.

International Organizations

United Nations and UN Specialized Agencies

UN and its specialized agencies play a pivotal role in adopting the Earthmate principles. These international bodies provide a platform for collaboration, knowledge sharing, and coordinated action on global environmental challenges. Through initiatives like UNFCCC, the UN addresses the interconnected nature of climate issues, promoting unity in efforts to mitigate climate change. However, the UN's bureaucratic structure and the challenge of reconciling diverse national interests can hinder the effectiveness of these initiatives. Striking a balance between sovereignty and global cooperation poses a challenge, and the need for unanimous decisions can sometimes impede swift action on urgent environmental matters.

Global Specialized Conservation and Biodiversity Organizations

Global specialized conservation and biodiversity organizations play a crucial role in embodying the Earthmate principles. These organizations, including CBD and IUCN, exemplify the positive impact of international collaboration on preserving Earth's biodiversity. They facilitate the exchange of knowledge, best practices, and resources among nations, promoting a shared responsibility for the stewardship of the planet. By fostering a sense of unity despite the diversity of ecosystems and species, these organizations contribute to the realization of the Earthmate principles. However, challenges such as the limited enforcement power of international agreements and the varying commitment levels of member states can hinder the effectiveness of these efforts, highlighting the nuanced interplay between global collaboration and national interests.

Non-environmental Organization and Environmental Issues

Several non-environmental organizations play a pivotal role in adopting the Earthmate principles, contributing to the philosophy's emphasis on unity despite plurality. WEF, for instance, brings together leaders from diverse sectors, fostering dialogue on global challenges, including environmental sustainability. Through collaborative efforts, these organizations integrate the Earthmate principles by recognizing the interconnectedness of economic prosperity, social well-being, and environmental protection. The Gates Foundation, primarily focused on health and poverty alleviation, invests in environmental initiatives, showcasing the integration of sustainability into broader development goals. The engagement of non-environmental organizations in the Earthmate principles reflects a holistic approach to addressing interconnected challenges.

International Environmental Legal Instruments

International environmental agreements, protocols, conventions, and treaties play a crucial role in advancing the Earthmate principles. These agreements serve as frameworks for global cooperation, aligning with Earthmate's emphasis on unity despite diversity. UNFCCC exemplifies this positive role by providing a platform for nations to collectively address climate challenges. The Paris Agreement, born out of UNFCCC, further emphasizes the interconnectedness of environmental, social, and economic aspects, aligning with Earthmate's holistic approach. These agreements contribute to fostering responsible stewardship of the planet, recognizing the shared responsibility for addressing global environmental issues. A potential negative effect of international environmental agreements, protocols, conventions, and treaties is that they often rely on voluntary commitments, which can lead to inconsistent implementation and lack of enforcement. Some countries may prioritize their economic interests over environmental obligations, leading to uneven progress. Additionally, the slow pace of negotiations and political disagreements among member nations can delay urgent action on critical environmental issues, hindering the effectiveness of these agreements in fully realizing the Earthmate principles of global stewardship and sustainability. This challenge can result in fragmented efforts and missed opportunities to address pressing environmental crises.

Global Environmental Funds and Financing Mechanisms

Global environmental funds and financing mechanisms play a crucial positive role in advancing the Earthmate principles. Initiatives like GCF, GEF, and GFDRR are designed to support climate mitigation and adaptation projects in developing countries. This reflects the Earthmate philosophy's emphasis on unity and collective responsibility, ensuring that nations, despite their diversity, have access to resources for addressing shared environmental challenges. These financial mechanisms align with Earthmate's call for responsible stewardship and recognize the interconnectedness of nations in addressing global environmental issues. A potential negative effect of global environmental funds and financing mechanisms is that the distribution of funds can be slow, bureaucratic, and inefficient, often hampering timely implementation of critical climate and environmental projects. Additionally, there is the risk of funds being misallocated or poorly monitored, leading to corruption or misuse. Some projects may not adequately consider the unique local needs or may prioritize short-term goals over long-term sustainability, which can undermine the effectiveness of the initiatives in fully realizing Earthmate's principles of responsible and equitable global stewardship. Furthermore, disparities in fund allocation between developed and developing nations could exacerbate inequalities, rather than fostering the unity Earthmate advocates for.

Humanitarian and Environmental Aid Organizations

Humanitarian and environmental aid organizations play a pivotal role in disaster response and preparedness, aligning with Earthmate philosophy's principles of unity despite plurality. Organizations like Médecins Sans Frontières (Doctors Without Borders) address the interconnectedness of humanitarian and environmental crises by providing crucial medical assistance during environmental disasters. This reflects Earthmate's emphasis on responsible stewardship, acknowledging the integral part humans play in nature. These organizations contribute to the collective well-being by responding to immediate needs during disasters, emphasizing the unity needed to navigate through such crises. Successful disaster response initiatives demonstrate the positive impact of organizations adopting the Earthmate principles, highlighting the interdependence of human and environmental well-being. A potential negative effect of humanitarian and environmental aid organizations is that their efforts can sometimes be hindered by logistical challenges, political instability, or inadequate coordination with local governments. This can result in delayed or inefficient aid delivery, undermining the immediate impact of their interventions. Additionally, without sufficient community involvement or understanding of local contexts, aid initiatives may inadvertently overlook sustainable long-term solutions, contradicting the Earthmate principles of responsible stewardship and unity. Furthermore, these organizations may face issues of donor fatigue, limiting their ability to consistently provide resources for both humanitarian and environmental crises over time.

International Research and Scientific Collaboration

International research and scientific collaboration play a crucial role in advancing the Earthmate principles by engaging in global environmental monitoring. Organizations such as the Belmont Forum and ICSU actively contribute to this endeavor. Their efforts align with Earthmate's emphasis on interconnectedness and responsible stewardship, as they work to collect and analyze data related to climate change, biodiversity loss, and pollution on a global scale. By fostering collaboration among scientists worldwide, these organizations contribute to a collective understanding of Earth's challenges, promoting a shared commitment to environmental well-being. A potential negative effect of international research and scientific collaboration is that despite their positive contributions, these efforts may be hampered by unequal access to resources, technology, and data among participating countries. Wealthier nations often dominate the research agenda, which can lead to a focus on issues that benefit them, while underfunded regions or topics crucial to less developed nations may be sidelined. This can create imbalances in global environmental solutions, undermining the equitable approach championed by the Earthmate principles. Additionally, differences in regulatory frameworks and intellectual property laws across countries can sometimes hinder the free exchange of scientific knowledge.

Special Environment-Oriented Entities

Standardization and Certification

Standardization and certification play a crucial role in advancing the Earthmate principles, particularly in the realm of carbon offset projects. Advocating for standardized methodologies ensures consistency and comparability across different projects, fostering trust in their environmental benefits. By endorsing third-party verification, the Earthmate philosophy's emphasis on responsible stewardship is upheld, as independent audits help ensure the credibility and effectiveness of emissions reduction initiatives. Encouraging providers to diversify their portfolio of carbon offset projects further aligns with Earthmate's holistic approach, addressing various environmental challenges such as reforestation, renewable energy, methane capture, and sustainable agriculture. Through these standardized practices, the Earthmate principles integrates seamlessly with efforts to combat climate change and promote responsible environmental practices.

Carbon Offset and Environmental Credit Providers

Carbon offset providers play a positive role in adopting the Earthmate principles by developing and promoting user-friendly carbon footprint calculators. These tools serve as educational instruments, raising public awareness about individual carbon footprints and encouraging engagement in carbon offsetting. This aligns with Earthmate's emphasis on education and transparency, as these initiatives empower individuals to make informed decisions about their environmental impact. By simplifying complex concepts related to carbon offsetting, providers contribute to the Earthmate vision of fostering global cooperation and addressing environmental challenges through accessible, public-friendly means.

Biodiversity Offsetting

Biodiversity offsetting, akin to carbon offsetting, can play a positive role in adopting the Earthmate principles, especially through conservation partnerships. Companies engaging in development projects that may impact biodiversity can invest in biodiversity offset programs, demonstrating a commitment to responsible stewardship. By establishing partnerships with conservation organizations, businesses can actively contribute to biodiversity conservation projects. This involvement may include funding research, habitat restoration, or wildlife protection initiatives. Such partnerships align with Earthmate's emphasis on interconnectedness and responsible environmental management, showcasing collaborative efforts between corporate entities and conservation organizations to address biodiversity concerns.

Innovations in Carbon Markets and Offset Financing: The Role of Blockchain Technology

Innovations in carbon markets, particularly through the integration of blockchain technology, are crucial for advancing the Earthmate principles. Blockchain's transparent and decentralized framework directly supports Earthmate's focus on interconnectedness and accountability. For instance, by utilizing blockchain, carbon markets can significantly enhance the traceability of carbon credit transactions, thereby ensuring their integrity. Each carbon credit can be represented as a unique digital token on the blockchain, allowing for real-time tracking of its lifecycle—from issuance to retirement.

A notable example is the partnership between IBM and Veridium, which aims to tokenize carbon credits on the blockchain, thus streamlining the verification process and reducing the potential for fraudigns with Earthmate's vision of using advanced technological solutions to navigate environmental complexities, fostering trust among stakeholders.

Furthermore, projects like AirCarbon are leveraging blockchain to create a global carbon trading platform, making it easier for companies to offset their carbon footprints. By ensuring that every transaction is immutable and publicly accessible, they promote transparency and collaboration within the carbon market.

Howevblockchain presents numerous benefits, it also introduces challenges, particularly regarding energy consumption. The mining process for certain blockchain networks can be energy-intensive, raising concerns about the environmental impact. For example, Bitcoin's network has faced criticism for its substantial carbon footprint. Thus, it's essecarbon market innovations to balance technological advancement with environmental stewardship, ensuring that solutions contribute positively to climate goals.

In conclusion, integrating blockchain technology into carbon markets exemplifies a proactive approach to promoting the Earthmate principles. By enhancing transparency and accountability while also addressing potential environmental drawbacks, stakeholders can work towards a more sustainable future.

PES

PES emerges as a positive force in aligning with the Earthmate principles by recognizing the intrinsic value of nature and fostering responsible stewardship. PES programs, whether initiated by governments or companies, establish frameworks where individuals or communities are incentivized for providing vital ecosystem services. This approach resonates with Earthmate's emphasis on interconnectedness and responsible engagement with nature. By rewarding actions that contribute to maintaining biodiversity-rich areas or practicing sustainable land use, PES programs actively promote a balance between human activities and ecological well-being.

Successful Practices and Initiatives

Individuals

Ordinary People

In scrutinizing successful practices and initiatives, the transformative influence of ordinary individuals becomes evident. We witness ordinary people effecting meaningful change by cultivating a sense of unity amidst the apparent diversity within their local communities. Whether through eco-friendly lifestyles and modest local endeavors or expansive global movements, ordinary individuals have proven their capacity to instigate meaningful change. Communities embracing eco-friendly practices, endorsing local businesses, and actively participating in conservation efforts exemplify the practical and successful application of the

Earthmate principles. Notably, local clean-up networks serve as exemplary models, illustrating how communities can collaboratively address environmental challenges and illustrate the potency of grassroots movements in effecting tangible improvements in environmental well-being. These Local Clean-Up Networks, as an example, not only highlight the efficacy of community-driven efforts in environmental conservation but also showcase the transformative power of collective action when individuals align with the Earthmate principles. This hands-on involvement fosters a deeper connection between individuals and their surroundings, reinforcing the Earthmate principles in a palpable and measurable manner. Spearheaded by ordinary citizens, these initiatives and success stories underscore the practical applicability of the Earthmate principles in real-world scenarios and offer tangible proof that these principles can be translated into practical action. Ordinary people, by engaging in such initiatives, emerge as vital contributors to the interconnected nexus of knowledge and action envisioned by the Earthmate principles. These instances and the positive outcomes of these practices not only underscore the potential for widespread adoption of the Earthmate principles at the grassroots level and the positive consequences of adopting the Earthmate principles but also stand as inspiring benchmarks for others, cultivating a collective sense of unity and responsibility for the planet.

Experts and Academia

Numerous items showcase the successful application of the Earthmate principles across diverse fields. Environmental engineers have implemented sustainable infrastructure projects that prioritize the unity of human development and ecological balance. In the realm of biodiversity research, initiatives centered on protecting and restoring ecosystems demonstrate tangible benefits to both humanity and nature. Renewable energy scientists have played a pivotal role in developing technologies that align with the Earthmate principles, fostering a harmonious coexistence with the environment. Environmental data visualization experts contribute by communicating complex ecological concepts to the broader public, fostering awareness and understanding. These items underscore the practicality and positive outcomes achievable through the integration of the Earthmate principles into various disciplines.

Environmental Writers, Filmmakers and Documentarians

Numerous success stories highlight the success of the Earthmate principles in the works of environmental animators, climate change documentarians, and advocacy-focused filmmakers. Environmental animators, through their creative and visually engaging productions, have successfully conveyed the Earthmate principles to diverse audiences, instigating positive changes in behavior and consciousness. Climate change documentarians, through their impactful storytelling, have contributed to the initiation of global conversations and policy changes. Advocacy-focused filmmakers, with their dedication to environmental causes, have played a vital role in mobilizing communities and inspiring collective action. The success of these initiatives showcases the transformative power of Earthmate when integrated into the creative expressions of these artists and communicators.

Celebrities, Influencers, Public Figures

Successful showcases abound within the realms of eco-friendly celebrities, sustainable fashion influencers, and environmental podcasters who have effectively integrated the Earthmate principles into their platforms. Eco-friendly celebrities, through endorsements and partnerships with environmental organizations, have directed significant resources towards sustainable initiatives, fostering positive change. Sustainable fashion influencers have demonstrated the feasibility of the Earthmate principles in the fashion industry, influencing both consumers and brands to adopt more environmentally conscious practices. Environmental podcasters, through engaging discussions and interviews, have brought the Earthmate principles to a broader audience, fostering a deeper understanding of interconnectedness. These cases underscore the potential for positive impact when influential individuals authentically embrace and champion the Earthmate principles.

Religious and Ethical Leaders

Illustrations abound within the realm of religious and ethical leaders who have successfully integrated the Earthmate principles into their teachings and practices. Green spiritual leaders, such as the Dalai Lama, have advocated for environmental conservation as an intrinsic part of their spiritual teachings, fostering a sense of interconnectedness with nature. Interfaith environmental advocates, like the Parliament of the World's Religions, have facilitated collaborative initiatives addressing environmental challenges from a global, multi-faith perspective. Ethical philosophers, such as Peter Singer, have contributed to ethical frameworks that emphasize the responsibility of individuals and societies towards the planet. These real-world examples illustrate the transformative potential of integrating the Earthmate principles into religious and ethical discourse, fostering a harmonious relationship between spiritual values and ecological awareness.

Environmental Reporters and Journalists

Successful inspiring instances within environmental journalism demonstrate the positive impact of adopting the Earthmate principles. Solutions-focused journalists have reported on communities implementing Earthmate-inspired initiatives, fostering a sense of hope and encouraging wider adoption. Environmental photojournalists have captured and disseminated powerful images that resonate with the interconnectedness central to Earthmate, fostering emotional connections and galvanizing support for environmental causes. Investigative environmental reporters have played a vital role in uncovering issues that require attention, aligning with Earthmate's emphasis on responsible stewardship. These cases illustrate the potential for environmental journalism to not only inform but also inspire positive change when rooted in the Earthmate principles.

Influential Individuals in Sustainable Business

Numerous examples showcase influential figures within various industries successfully integrating the Earthmate principles into their sustainable business practices. Eco-friendly designers, for instance, have created fashion lines that prioritize sustainable materials and ethical production methods, promoting Earthmate's interconnectedness in the fashion industry. Sustainable tech innovators have developed products with reduced environmental impact, showcasing how technological advancements can coexist responsibly with the planet. Environmental monitoring innovators have utilized technology to gather data on ecological systems, contributing to a better understanding of the Earthmate principles in action. These cases highlight the tangible successes achieved when influential individuals embrace and implement the Earthmate principles within their respective fields.

Environmental Activists and Advocates

Successful empowering examples abound within various subcategories of environmental activism, showcasing the positive impact of the Earthmate principles in action. Conservation campaigners, such as those involved in reforestation initiatives, demonstrate how the Earthmate principles guide efforts to restore and maintain the balance of ecosystems. Affected community advocates, when successful, contribute to a harmonious coexistence between human communities and the environment, addressing environmental injustices. Policy advocates for environmental justice, through legal and systemic changes, establish frameworks aligning with Earthmate values. Youth activists, like participants in the School Strike for Climate, youth environmental hacktivists, and attendees of youth climate conferences, bring fresh perspectives and innovative solutions to the forefront, embodying the Earthmate principles in their activism.

Environmental Justice and Climate Justice Advocates

Successful inspiring instances within environmental and climate justice activism showcase the positive impact of the Earthmate principles in addressing systemic inequalities. Environmental racism activists, through their efforts, have contributed to legal and policy changes that address both social justice and environmental concerns. Gender and environment advocates have successfully integrated the Earthmate principles into their campaigns, fostering awareness of the interconnectedness between gender issues and environmental degradation. Indigenous rights and land defenders, through legal battles and grassroots movements, exemplify the Earthmate principles by emphasizing responsible stewardship of the planet, acknowledging the integral relationship between humans and nature.

Culinary Experts and Food Influencers and Their Recipes

Successful practical instances within sustainable cooking channels exemplify the possible positive impact of the Earthmate principles in culinary practices. Influencers and chefs adopting these principles showcase recipes that prioritize local, seasonal produce, reducing the carbon footprint associated with food transportation. These platforms also provide educational content, fostering awareness about the interconnectedness of food choices and ecological well-being. By integrating Earthmate values into their culinary initiatives, these influencers contribute to a broader shift in consumer behavior, encouraging sustainable and responsible food practices.

Traditional Knowledge Holders

Successful inspiring instances within indigenous communities highlight the positive impact of incorporating traditional knowledge into the Earthmate principles. Indigenous elders and wisdom keepers in various regions have successfully advocated for land conservation, recognizing the intricate ties between cultural heritage and ecological well-being. Traditional herbalists have been instrumental in promoting the sustainable harvesting of medicinal plants, contributing to biodiversity conservation. Indigenous artisans, by integrating traditional practices with sustainable materials, showcase how cultural expressions can be in harmony with Earthmate values. These cases underscore the potential for traditional knowledge to inform and enrich environmental conservation efforts while maintaining cultural integrity.

Educational and Consultative Institutions

Schools and Environmental Education Programs

Successful inspiring instances within schools and environmental education programs showcase the possible positive impact of adopting the Earthmate principles. Schools with innovative environmental education curricula, emphasizing hands-on learning experiences such as school gardens or nature-based activities, provide students with tangible connections to the environment. Nature-based education programs, specifically designed for experiential learning in natural environments, offer students a deeper understanding of their interconnectedness with the natural world. Carbon literacy programs within schools help students and businesses make informed decisions about offsetting their carbon footprint, fostering a sense of responsibility. Transparency in offset education initiatives ensures that students and businesses are well-informed about the limitations, benefits, and considerations of carbon offsets, promoting responsible environmental practices.

Colleges and Universities and Environmental Education Programs

Successful empowering examples within colleges and universities highlight the possible positive impact of adopting the Earthmate principles. Institutions that have implemented sustainable practices on campus showcase leadership in environmental stewardship. This includes initiatives like renewable energy adoption, waste reduction programs, and eco-friendly campus designs. Distinguishing between undergraduate and graduate programs, emphasizing specialized environmental disciplines, allows for a tailored approach to the Earthmate principles based on academic levels. Recognizing institutions known for groundbreaking environmental research and interdisciplinary studies underscores the impact of higher education in pushing the boundaries of knowledge. Acknowledging partnerships between universities and external organizations for applied environmental research exemplifies how academic institutions can actively contribute to practical solutions for environmental challenges.

Informal Online or Offline Environmental Education

Successful inspiring instances in informal environmental education highlight initiatives that effectively leverage digital technologies and online platforms. Recognizing institutions and programs targeting younger demographics showcases efforts to instill environmental awareness from an early age. Acknowledging institutions offering certification programs for professionals in sustainability management and green practices emphasizes the practical application of the Earthmate principles in various fields. The role of online educational platforms and MOOCs in disseminating environmental knowledge globally underscores the impact of technology in reaching diverse audiences. Environmental non-profit education programs and initiatives exemplify how organizations actively engage in raising awareness and fostering understanding of environmental issues among schools, communities, and marginalized groups.

Partnerships and Collaborations Targeting Environmental Education

Successful results illustrate the impact of partnerships and collaborations in environmental education. Initiatives that specifically incorporate traditional ecological knowledge showcase the successful integration of indigenous wisdom into mainstream education, enriching the understanding of interconnected systems. Community-based environmental education programs, facilitated through collaborations, emphasize practical knowledge and skills, empowering local communities to address environmental challenges effectively. Collaboration between educational institutions and businesses not only contributes to practical solutions but also establishes a model for mutually beneficial relationships that prioritize environmental sustainability.

Environmental Consultants and Advisory Services

Successful results illustrate the positive impact of environmental consultants in implementing Earthmate principles. Projects focusing on environmental impact assessments showcase how consultants contribute to preventing and mitigating potential ecological harm. Additionally, initiatives by consultants specializing in pollution control demonstrate effective measures to address environmental fragility. Collaborations between consultants and businesses, governments, or NGOs provide real-world examples of how the Earthmate principles can be integrated into diverse sectors. These cases emphasize the importance of interdisciplinary collaboration and the practical application of scientific actions for a sustainable future.

Think Tanks and Policy Institutes

Successful proven strategies demonstrate the positive impact of think tanks and policy institutes in implementing the Earthmate principles. For instance, think tanks focusing on climate policy have successfully influenced the development and adoption of environmentally sustainable regulations. Initiatives that translate research into actionable policy recommendations showcase the practical application of the Earthmate principles in shaping real-world policies. Recognizing think tanks that have contributed to the development of international agreements on climate change highlights the global reach and effectiveness of Earthmate-oriented policies.

R&D Institutions

Successful proven strategies showcase the positive impact of R&D institutions in implementing the Earthmate principles. For example, research centers focusing on climate modeling contribute significantly to understanding and mitigating the impacts of climate change. Institutions dedicated to renewable energy development play a crucial role in transitioning towards sustainable and clean energy sources. R&D initiatives in sustainable agriculture demonstrate how technological advancements can promote responsible resource consumption and regenerative agricultural practices. Highlighting these cases emphasizes the practical application of the Earthmate principles through the work of R&D institutions.

Media and Culture

Media and Communication

Successful showcases exemplify the positive impact of media outlets in implementing Earthmate practices. For instance, independent media sources focusing on environmental issues contribute to a more diverse and nuanced understanding of ecological challenges. Social media platforms provide a space for grassroots movements and environmental campaigns to gain momentum. Environmental documentaries, films, and storytelling platforms showcase successful initiatives and positive environmental actions, influencing public perceptions positively. Recognizing and highlighting such cases emphasizes the crucial role media plays in advocating for a sustainable and ecologically balanced future.

Social Media and Online Activism

Numerous success stories illustrate the successful integration of the Earthmate principles through social media and online activism. Global citizens and networks have effectively utilized these platforms to advocate for environmental causes on an international scale. Initiatives that foster international collaboration showcase the potential for online activism to transcend geographical boundaries. The active engagement of global citizens and youth in environmental action exemplifies the transformative impact of online platforms. Non-profit organizations, leveraging social media for awareness campaigns, have demonstrated the ability to amplify their messages and mobilize public support. These cases collectively highlight the positive role of social media and online activism in advancing the Earthmate principles and fostering a collective commitment to environmental stewardship.

Religious and Ethical Perspectives

Numerous inspiring instances highlight the successful integration of the Earthmate principles through religious and ethical perspectives. Religious communities often engage in environmental initiatives inspired by their teachings. For instance, the Green Church Movement, where Christian congregations undertake eco-friendly practices, or Buddhist monasteries promoting conservation reflect a commitment to the Earthmate principles. Ethical organizations, guided by principles like eco-philosophy, contribute to sustainable development by aligning their practices with ecological balance. Indigenous communities, driven by their cultural and spiritual connections to the land, have demonstrated successful practices by incorporating traditional ecological knowledge into modern conservation efforts.

Art and Literature

Numerous impact narratives exemplify successful Earthmate practices and initiatives within the realms of art and literature. Environmental art installations, such as Chris Jordan's photographic series depicting the impact of plastic waste, have effectively raised awareness about consumption patterns. The fashion industry has seen success stories, like Patagonia's commitment to sustainability, implementing fair labor practices and promoting responsible consumerism. Interface, a global carpet manufacturer, is another success story that pioneered sustainable business practices with its Mission Zero commitment to eliminate negative environmental impacts by 2020. Cultural events, exhibitions, and performances, such as the Earth Day Network's Arts for the Earth, bring together artists and audiences to celebrate environmental themes, fostering a collective commitment to stewardship.

Profit-Oriented Entities

Financial Institutions

Several financial institutions have successfully integrated the Earthmate principles into their practices, serving as exemplars for the industry. Impact investing has gained prominence, directing capital towards companies generating positive social and environmental outcomes alongside financial returns. Institutions actively participating in green banking, offering financial products with a focus on sustainability, contribute significantly. Initiatives like green bonds, where funds raised are dedicated to environmentally beneficial projects, exemplify successful Earthmate practices within financial institutions.

Industrial Productors

Numerous industrial producers exemplify successful Earthmate practices and initiatives, showcasing a commitment to sustainability. Companies embracing circular economy principles to minimize waste, reduce carbon footprints, and adopt sustainable supply chain practices serve as proof points. Entities conducting LCA to understand and minimize environmental impact, coupled with robust CSR programs extending beyond compliance, contribute significantly to Earthmate goals. Industries implementing biodiversity conservation measures, such as habitat restoration and preservation, exhibit successful initiatives contributing to the holistic vision of environmental stewardship.

Mining Industry

Several practical applications exemplify successful Earthmate practices within the mining industry. Some companies actively engage in sustainable mining, emphasizing land reclamation as a fundamental aspect of their operations. Beyond regulatory compliance, these companies undertake additional measures to restore ecosystems and promote biodiversity in the reclaimed areas. Initiatives that focus on responsible sourcing showcase efforts to minimize the environmental and social footprint associated with mining activities. Furthermore, companies promoting the use of recycled materials within the mining industry contribute to a circular economy, aligning with the Earthmate principles by reducing the demand for new raw materials and mitigating environmental impact.

Agriculture, Agroforestry, Permaculture, Sustainable Agriculture and Forestry Initiatives

Successful Earthmate practices within the realm of agriculture and forestry include companies that champion organic farming and biodiversity conservation. These entities prioritize agroecological approaches, emphasizing the importance of sustainable land use. Recognizing and showcasing companies that actively engage in regenerative agriculture contributes to a growing body of success stories. Agroforestry practices, when integrated into carbon offset projects, provide a compelling example of sustainable land use that not only sequesters carbon but also supports agricultural production. Certification standards for sustainable agriculture and forestry practices further contribute to the success stories, providing clear benchmarks for environmentally responsible initiatives.

Technological Innovators and Environmental Technology Startups

Examining successful Earthmate practices in the technological sector reveals inspiring instances. Companies specializing in CCS demonstrate a commitment to offsetting emissions from industries with limited alternative solutions. Their innovative approaches contribute to mitigating climate change by capturing and storing carbon dioxide, reducing GHG emissions. Another notable case involves advancements in DAC technologies. Startups focusing on these solutions actively engage in removing carbon dioxide directly from the atmosphere, offering a pathway to achieve negative emissions. These cases highlight how technology-driven initiatives can lead to tangible and positive environmental outcomes.

Tourism Industry, Sustainable Tourism and Ecotourism

Examining successful Earthmate practices within the tourism industry reveals impactful initiatives. Organizations promoting responsible tourism practices combine conservation efforts with economic opportunities for local communities. These initiatives prioritize preferential procurement, ensuring that businesses support products and services from companies with strong environmental practices. Additionally, ecotourism principles, which emphasize conservation and community involvement, contribute positively to Earthmate goals. Successful projects actively engaging indigenous and local communities showcase how tourism can be a force for environmental protection and community well-being simultaneously.

In line with this, the recent rebranding of WTO to "UN Tourism" aligns with its commitment to sustainability and responsible tourism. This change reflects an emphasis on integrating environmental goals more thoroughly into global tourism practices, focusing on the dual objectives of economic benefit and environmental protection. Through its redefined mission, UN Tourism aims to drive sustainable tourism initiatives that prioritize both conservation and support for local communities. This new direction underscores the agency's focus on using tourism as a tool for achieving broader SDGs, especially in fostering sustainable practices and reducing environmental impact worldwide. By renaming and rebranding itself, UN Tourism seeks to strengthen its leadership in eco-friendly tourism policies and establish itself as a key player in creating environmentally responsible tourism practices globally.

Supply Chains and Consumer Behavior

Several tangible results showcase successful Earthmate practices within supply chains. Companies transforming their supply chains to prioritize sustainability, ethical sourcing, and fair labor practices demonstrate the feasibility of aligning business operations with the Earthmate principles. Initiatives engaging consumers in making sustainable choices and contributing to the circular economy demonstrate the potential impact of supply chain practices on consumer behavior. Furthermore, the adoption of blockchain and other technologies for traceability and transparency in supply chains has been a successful strategy in ensuring sustainable sourcing. These cases highlight the transformative power of supply chains in driving positive environmental and social outcomes.

Green Building and Construction

Numerous project profiles highlight successful Earthmate practices within the realm of green building and construction. Companies that have prioritized sustainable building practices and achieved notable success in energy efficiency provide tangible examples of how the Earthmate principles can be applied in the construction sector. Projects showcasing the use of eco-friendly materials, innovative design for environmental conservation, and the integration of renewable energy sources serve as inspirations for the wider industry. These cases underscore the feasibility of adopting the Earthmate principles in construction and emphasize the positive impact such practices can have on both the environment and society.

Packaging and Sustainable Packaging Companies

Successful practices in sustainable packaging provide tangible examples of how companies have aligned their operations with the Earthmate principles. Businesses that have transitioned to innovative and eco-friendly packaging materials, such as biodegradable plastics, compostable materials, and recyclable packaging, demonstrate the feasibility of adopting

Earthmate-aligned practices in the packaging industry. These cases highlight the positive impact that sustainable packaging initiatives can have on reducing environmental degradation and promoting responsible resource consumption. Successful examples inspire other companies to follow suit, fostering a collective commitment to the Earthmate principles within the packaging sector.

Waste Management Industry

Impactful initiatives of successful practices within the waste management industry provide tangible examples of how companies can make a positive impact. Initiatives that focus on recycling technologies, waste-to-energy solutions, and sustainable waste practices showcase real-world applications of the Earthmate principles. For instance, companies implementing advanced recycling technologies, such as chemical recycling, demonstrate innovative approaches to address the complexities of waste management. Successful cases highlight the feasibility and effectiveness of adopting Earthmate-aligned practices, inspiring other waste management entities and businesses across various sectors to follow suit.

Renewable Energy Industry

Examining in-depth initiatives of successful practices within the renewable energy industry provides tangible examples of its positive impact. Initiatives that support innovative renewable energy technologies showcase how scientific principles can be leveraged for the benefit of the planet. Successful projects highlight the feasibility and effectiveness of adopting Earthmate-aligned practices, inspiring further advancements in clean energy production. By spotlighting these cases, the renewable energy sector contributes to Earthmate's vision of a sustainable and ecologically balanced future. Such success stories serve as beacons, guiding other industries and sectors toward environmentally conscious practices and fostering a collective commitment to address global challenges.

Public–Private Partnerships

Exploring implementations of successful practices facilitated by PPPs provides tangible evidence of their positive impact. Instances where profit-oriented entities collaborated with non-profit organizations or research institutions to drive sustainable innovation showcase the potential of such partnerships. Successful projects underscore the effectiveness of combining economic prosperity, social well-being, and environmental protection—a key tenet of the Earthmate principles. By recognizing and highlighting these partnerships, Earthmate aims to inspire similar collaborations across different sectors. Through PPPs, tangible contributions to Earthmate's vision of a sustainable and ecologically balanced future become evident, demonstrating the transformative power of unified efforts.

Non-profit Entities

Environmental Justice and Climate Justice Associations

Examining strategies of successful practices by Environmental Justice and Climate Justice Associations provides tangible evidence of their positive impact on the Earthmate principles. Organizations that focus on policy advocacy and legal support to influence environmental legislation showcase the practical implementation of Earthmate's holistic approach. These efforts underline the interconnected nature of scientific, political, and societal realms, addressing Earth's environmental crises through interdisciplinary collaboration. By navigating the complexities of our time and fostering global cooperation, these associations contribute to Earthmate's vision of a sustainable and ecologically balanced future. Successful initiatives highlight the potential for non-profits to influence legal frameworks positively, promoting unity amid the diversity of interests and priorities.

Wildlife Conservation Associations

Examining strategies of successful wildlife conservation practices provides tangible evidence of positive outcomes aligned with the Earthmate principles. Associations engaged in habitat preservation, species recovery, and ecosystem restoration showcase the application of scientific principles rooted in Earthmate's holistic approach. These initiatives bridge understanding gaps, foster global cooperation, and address cultural sensitivities by recognizing the interplay between human communities and the natural environment. By embracing circular economy principles and advocating for responsible resource consumption, these associations contribute to Earthmate's call for a delicate balance between economic prosperity, social well-being, and environmental protection.

Climate Change Mitigation and Adaptation Associations

While global success in this field has yet to be fully realized, delving into strategies of climate change mitigation and adaptation practices reveals promising instances. Associations that implement innovative strategies, addressing both the mitigation of climate change impacts and assisting communities in adapting to shifting conditions, stand as concrete examples of Earthmate's plea for interdisciplinary collaboration. These instances showcase the potential for successful and impactful initiatives, contributing to Earthmate's vision of responsible stewardship and interconnected solutions. These initiatives underscore the importance of partnerships with diverse sectors, such as technology, policy, and community development, to address the complex challenges posed by climate change. By embracing circular economy principles and advocating for responsible resource consumption, these associations contribute to Earthmate's vision of a sustainable and ecologically balanced future.

Ocean Protection Associations

Several Ocean Protection Associations serve as real-world examples for successful practices, showcasing tangible positive outcomes. These associations dedicate themselves to marine conservation, actively combating plastic pollution, overfishing, and championing the establishment of marine protected areas. By undertaking such initiatives, these organizations contribute significantly to Earthmate's vision of a sustainable and ecologically balanced future. Moreover, those organizations engaged in ocean research, promoting marine biodiversity, and addressing the impacts of climate change on marine ecosystems demonstrate the practical application of Earthmate's call for interdisciplinary collaboration and holistic environmental protection.

Community-Based Conservation Initiatives

Examining narratives of successful practices within Community-Based Conservation Initiatives reveals tangible outcomes that align with the Earthmate principles. These initiatives emphasize partnerships with indigenous and local communities, integrating traditional knowledge into conservation practices. By doing so, they bridge understanding gaps and address cultural sensitivities, creating a more inclusive and holistic approach to environmental protection. Additionally, initiatives that link conservation efforts with sustainable livelihood opportunities for local communities showcase Earthmate's commitment to a delicate balance between economic prosperity, social well-being, and environmental protection.

Green Philanthropy and Impact Investing

Examining implementations of successful practices within the realm of green philanthropy and impact investing reveals tangible outcomes that align with the Earthmate principles. Successful initiatives showcase how these funding models can bridge understanding gaps, foster global cooperation, and address cultural sensitivities integral to Earthmate's call to action. For instance, organizations involved in green philanthropy may support projects that promote responsible resource consumption and circular economy principles. Impact investing can facilitate the implementation of sustainable development principles, fostering regenerative approaches to production. These cases demonstrate the potential for financial mechanisms to drive positive environmental change in alignment with the Earthmate principles.

Legal Advocacy and Environmental Law NGOs

Examining results of successful practices within the realm of legal advocacy and Environmental Law NGOs illustrates tangible outcomes that align with the Earthmate principles. Successful initiatives showcase how legal actions can prevent environmental degradation and promote responsible resource consumption. For example, legal efforts may lead to the strengthening of environmental laws and regulations, creating a framework for sustainable development. Collaborations between environmental-focused NGOs and those advocating for indigenous rights represent a holistic approach, recognizing the intertwined nature of environmental and social concerns. These cases exemplify the potential for legal advocacy to foster a world where humanity collectively commits to a sustainable and ecologically balanced future.

Crisis Response and Disaster Relief NGOs

Examining instances of successful practices within the domain of Crisis Response and Disaster Relief NGOs sheds light on tangible outcomes that align with the Earthmate principles. Successful initiatives demonstrate the positive impact of these organizations in mitigating the consequences of environmental disasters. For instance, rapid response and relief efforts during hurricanes, floods, and wildfires showcase the NGOs' ability to foster resilience and aid in the recovery of affected regions. These cases illustrate the potential for coordinated, collective action, emphasizing the interconnected nature of environmental challenges and the need for a unified approach to crisis response.

Governmental Entities

Legislative Bodies

In-depth explorations of successful practices within legislative bodies provides insight into tangible outcomes aligned with the Earthmate principles. Legislative efforts focused on the restoration of degraded ecosystems and climate change mitigation exemplify the holistic approach advocated by Earthmate. Instances of legislative bodies actively involved in creating and updating environmental laws and regulations underscore the commitment to responsible stewardship. Successful initiatives, such as environmental taxes, subsidies for sustainable practices, and cap-and-trade systems, demonstrate the positive impact of legislative actions in fostering a balance between economic prosperity, social well-being, and environmental protection.

Legal and Regulatory Frameworks

In-depth explorations of successful practices within legal and regulatory frameworks provides valuable insights into tangible outcomes aligned with the Earthmate principles and science. Instances where legal frameworks have been effective in preventing environmental degradation, improving air and water quality, and promoting sustainable waste management underscore the positive impact of these practices. Governments that regularly assess and update regulations to address emerging environmental challenges demonstrate a commitment to the Earthmate principles. Successful initiatives within legal frameworks serve as inspirations for other regions, encouraging the adoption of similar comprehensive approaches to environmental governance.

Law Enforcement Entities

In-depth explorations of successful practices within law enforcement entities provides tangible examples of positive outcomes aligned with the Earthmate principles. For instance, governments embracing E-government practices contribute positively to environmental sustainability. By transitioning from traditional paper-based processes to digital platforms, there is a significant reduction in paper consumption, leading to a decrease in deforestation and the environmental impact associated with paper production. Furthermore, E-government streamlines administrative processes, reducing the need for physical transportation and storage of documents, which results in lower carbon emissions. Digital communication channels employed in E-government enhance public engagement in environmental initiatives, fostering a sense of shared responsibility. Citizens can access environmental information, participate in public consultations, and engage with government agencies online, promoting transparency and collaboration. As a successful practice, E-government aligns with the principles of responsible resource consumption, innovation in sustainable practices, and fostering a connection between the government and the public to collectively address environmental challenges. In the next step, I anticipate witnessing the "Automated Governance System" combatting corruption, ensuring transparency, eliminating bias, and mitigating individual and party interests, ushering in a novel era of governance. This transformative approach, exemplified by terms such as Automated Executive, Robo-Governor, Algorithmic Regulator, Cybernetic Administrator, Digital Steward, Techno-Diplomat, Intelligent Governance Officer, Digital Policy Director, Virtual Governance Specialist, AI Governor, and possibly even Robotic Policy Maker, AI Political Strategist, Virtual Chief Executive, Autonomous Political Controller, Digital Leader, Cyber Sovereign, AI Statesperson, Quantum Leader, Synthetic Leadership, Neural Network Regent, has the potential to supersede traditional systems that often act as obstacles to embracing the Earthmate principles. Additionally, Successful initiatives involving local communities, Institutions and companies in monitoring, assessment, conservation and crime prevention highlight the importance of community-led environmental stewardship. These items serve as models for other regions, illustrating the potential of collaborative efforts between law enforcement, governments, and communities to achieve sustainable environmental outcomes. Recognizing and sharing such success stories is crucial for inspiring further initiatives and building momentum for Earthmate practices.

Judiciary

Landmark decisions in environmental litigation contribute significantly to the success of Earthmate practices. When judiciaries deliver rulings that set legal precedents in favor of environmental protection, it establishes a framework for future cases. A notable example is the case of Massachusetts v. EPA in the United States, where the Supreme Court recognized the EPA's authority to regulate GHG emissions. Such decisions serve as case studies for successful initiatives, demonstrating the judiciary's role in interpreting and enforcing laws that align with principles of responsible stewardship and sustainable development.

Governmental Organizations for Wildlife Conservation

Efforts to combat wildlife trafficking showcase successful practices where governmental organizations collaborate with law enforcement and international partners. Operation Thunderstorm, a collaborative initiative led by INTERPOL, successfully targeted illegal wildlife trade globally. Governmental agencies, such as customs and border protection, worked together to intercept traffickers, highlighting the positive impact of coordinated efforts. Such initiatives align with the Earthmate principles by addressing the socio-economic, political, and environmental factors contributing to wildlife exploitation and trafficking.

Government Water Resource Management Entities

A notable practice in water resource management is the adoption of IWRM principles by governmental entities. In South Africa, the National Water Act embraces IWRM to ensure sustainable and equitable water use. This initiative has successfully integrated the needs of ecosystems and communities, promoting a balanced and holistic approach to water resource management. Additionally, acknowledging water resource management entities that integrate climate resilience considerations into their planning and policies reflects a commitment to the Earthmate principles by addressing the interconnected challenges of water management and climate change.

Governmental Institutions for Meteorology, Climate Science, and Weather Forecasting

Institutions specializing in climate science and weather forecasting contribute significantly to successful practices. ECMWF exemplifies a case where international collaboration enhances the understanding of weather patterns and climate. The ECMWF's commitment to open data sharing and global cooperation aligns with Earthmate's emphasis on fostering global collaboration to address environmental challenges. Such initiatives contribute to bridging understanding gaps and addressing cultural sensitivities, promoting a holistic approach to Earth's interconnected nature.

Environmental Government Agencies

Numerous environmental government agencies worldwide have implemented successful practices and initiatives. SEPA stands out for its comprehensive approach to sustainability. SEPA integrates the Earthmate principles by prioritizing transparency in decision-making and actively engaging the public in environmental policy discussions. The agency has effectively collaborated with other government entities, demonstrating that success in environmental stewardship often requires inter-agency cooperation. This case showcases how adopting the Earthmate principles in government agencies fosters collective responsibility, transparency, and cross-sectoral collaboration for sustainable environmental practices.

Urban Planning, Green Infrastructure and Sustainable Development

Successful practices and initiatives in urban planning are exemplified by cities that prioritize sustainable development and resilience planning. Singapore stands out as a case study in urban planning and sustainable urban development. The city-state has implemented comprehensive strategies to address environmental challenges in its urban setting. Singapore's commitment to green architecture, extensive green spaces, and water management systems reflects an Earthmate-inspired approach to urban development. By integrating scientific principles into urban planning, Singapore demonstrates how the Earthmate principles can be translated into successful, tangible practices, creating a harmonious balance between urbanization and environmental conservation.

Health and Environment

Healthcare Industry

Several healthcare facilities serve as successful real-world examples in incorporating Earthmate principles. Institutions emphasizing energy-efficient operations, waste reduction, and sustainable procurement showcase tangible commitments to the principles. These initiatives demonstrate that sustainable practices are achievable in the healthcare sector, highlighting the importance of eco-friendly infrastructure and waste management. Recognizing programs that focus on biodiversity conservation and preparing for the health impacts of climate change exemplifies a holistic approach to healthcare. Additionally, initiatives addressing zoonotic diseases emphasize the interconnectedness of human, animal, and environmental health, showcasing the effectiveness of a "One Health" approach in healthcare systems worldwide.

Medical Infrastructure and Health Systems

Successful practices and initiatives within medical infrastructure and health systems emphasize the importance of addressing health disparities, promoting climate resilience, and integrating green infrastructure. Initiatives focusing on building climate-resilient healthcare infrastructure, ensuring hospitals can withstand and respond to the impacts of extreme weather events, showcase a forward-thinking approach aligned with the Earthmate principles. The role of green infrastructure, such as parks and urban green spaces, in promoting physical activity, mental well-being, and overall community health serves as another noteworthy case study. Urban planning strategies that ensure equitable access to green spaces further demonstrate a holistic approach to healthcare, emphasizing the positive impact of nature on both physical and mental health.

Environmental Psychology

Inspiring instances showcasing successful practices and initiatives within environmental psychology demonstrate the transformative potential of integrating psychological principles into environmental planning. Recognizing and promoting the positive effects of nature on mental health, some initiatives have successfully integrated nature-based interventions in healthcare settings. These practices align with Earthmate's call for a delicate balance between economic prosperity, social well-being, and environmental protection. Additionally, research on environmental factors affecting community well-being—such as green space access (Ma et al. 2019; Guite et al. 2006; Jabbar et al. 2022; Reyes-Riveros et al. 2021), air quality (Yuan et al. 2018; Lal et al. 2020; Guo et al. 2021), and noise pollution (Gidlöf-Gunnarsson and Öhrström 2007)—supports evidence-based strategies for building healthier, more sustainable communities. These studies highlight how various aspects of the environment contribute to overall quality of life, underscoring the importance of targeted interventions in urban planning and public health.

Community-Based Health Initiatives

Several inspiring instances illustrate successful community-based health initiatives that effectively integrate the Earthmate principles. These initiatives go beyond conventional health approaches, recognizing the interconnectedness of environmental and public health. Successful projects often involve grassroots efforts that engage communities in monitoring and addressing environmental issues, contributing to the creation of healthier and more sustainable living environments. By focusing on the well-being of both individuals and the environment, these initiatives exemplify the Earthmate principles in action, showcasing the positive impact of community-driven approaches.

Global Health and Environmental International Organizations

Organizations like HEAL provide successful practices and initiatives within the global health and environmental sector. By emphasizing prevention, conservation, and adaptation, HEAL exemplifies how international organizations can drive positive change. Successful initiatives often involve advocacy for policies that address the root causes of environmental and health issues, showcasing the interconnected nature of these challenges. HEAL's work, for instance, includes campaigns to reduce exposure to harmful chemicals, combat air pollution, and promote a transition to renewable energy. These cases illustrate the potential impact of aligning global health and environmental efforts with the Earthmate principles, contributing to a more sustainable and balanced future.

International Organizations

United Nations and UN Specialized Agencies

The UN has witnessed successful practices through various agencies and initiatives. UNEP has been instrumental in promoting sustainable development principles, advocating for the harmonious integration of economic, social, and environmental considerations. Additionally, IPCC has played a crucial role in synthesizing scientific knowledge, fostering global cooperation in addressing climate change. GCF stands out as an initiative supporting climate-related projects in developing countries, exemplifying collective efforts for a sustainable future. These items showcase the potential for international collaboration to address Earth's challenges, emphasizing the positive impact of coordinated action.

Global Specialized Conservation and Biodiversity Organizations

Numerous tangible results illustrate successful practices and initiatives led by global specialized conservation and biodiversity organizations. The IRENA and REN21 showcase the positive role these organizations play in promoting sustainable energy solutions. Their efforts contribute to mitigating climate change, aligning with the Earthmate principles' emphasis on responsible resource consumption. Additionally, the INECE exemplifies successful initiatives in enforcing environmental laws globally. These cases demonstrate the potential for international organizations to drive positive change through collaborative initiatives that address both ecological and social dimensions.

Non-environmental Organization and Environmental Issues

The Gates Foundation serves as a compelling case study of a non-environmental organization successfully implementing Earthmate practices. By investing in agricultural and environmental initiatives, the foundation addresses the interconnectedness of human well-being and environmental health. Similarly, the WEF, through its Sustainable Development Impact Summit, facilitates discussions and initiatives aligned with the Earthmate principles. These cases demonstrate that non-environmental organizations can leverage their influence, resources, and partnerships to drive positive change and contribute to the Earthmate vision of a balanced and interconnected world.

International Environmental Legal Instruments

The Earth Summit, organized by the UNCED, stands as a case study of a successful practice. The summit diffused key principles into subsequent environmental treaties, including the precautionary principle, common but differentiated responsibilities, and the polluter pays principle. The Kyoto Protocol, an international agreement aimed at mitigating climate change, exemplifies collaborative efforts guided by the Earthmate principles. By setting binding emissions targets for developed nations, the protocol acknowledges the interconnected responsibilities of nations in combating global environmental challenges.

Global Environmental Funds and Financing Mechanisms

The GCF serves as a notable case study, promoting a holistic approach in addressing climate challenges. By providing financial assistance to developing countries, the GCF supports projects that integrate environmental, social, and economic considerations, aligning with the Earthmate principles. The GEF, through its support for biodiversity conservation and sustainable development, exemplifies the interconnectedness of ecological and human systems. These funding mechanisms contribute to Earthmate's vision of a balanced and interconnected world by addressing environmental challenges through collaborative, cross-cutting projects.

Humanitarian and Environmental Aid Organizations

Amnesty International's initiatives addressing the intersection of environmental issues and human rights exemplify successful practices. By advocating for environmental justice and defending communities affected by environmental degradation, Amnesty International embraces the holistic approach promoted by the Earthmate principles. EarthRights International is another case study, actively working at the intersection of human rights and environmental protection. Their efforts align with Earthmate's call for responsible stewardship and recognition of the interconnected nature of ecological and social systems. These organizations contribute to a sustainable and ecologically balanced future by addressing both human and environmental needs, showcasing successful models for the integration of the Earthmate principles.

International Research and Scientific Collaboration

An exemplary case of successful practices is the establishment of open data platforms for sharing environmental research findings. Initiatives like the former ESSP, IGBP, and Diversitas have played a pivotal role in creating transparent and accessible repositories of environmental data. These platforms exemplify Earthmate's call for global cooperation and data transparency, providing a foundation for researchers, policymakers, and the public to access valuable information. Through open data initiatives, successful practices emerge as instruments of the Earthmate principles, emphasizing collaboration, shared knowledge, and a commitment to sustainable development.

Special Environment-Oriented Entities

Standardization and Certification

The implementation of eco-labeling and certification programs serves as a successful practice within the standardization framework. These programs verify the sustainability of products and services, empowering consumers to make informed choices aligned with the Earthmate principles. By incorporating sustainable development principles and embracing circular economy concepts, eco-labeling initiatives contribute to responsible resource consumption and a regenerative approach to production. Successful programs in this realm demonstrate the positive impact of standardization and certification in promoting environmentally conscious consumer behavior, fostering a collective commitment to a sustainable and ecologically balanced future.

Carbon Offset and Environmental Credit Providers

A successful practice within the carbon offset industry involves the avoidance of greenwashing. Greenwashing, the misleading representation of environmental efforts, is mitigated through the advocacy for clear standards and guidelines. By promoting honest and accurate representation of the environmental impact of offset projects, successful initiatives uphold the Earthmate principles of unity despite the plurality of actions. Companies committed to preventing greenwashing contribute to building a trustworthy framework in which consumers can confidently engage in carbon offsetting, fostering a genuine sense of environmental responsibility.

Biodiversity Offsetting

Successful practices within biodiversity offsetting often involve ecosystem restoration initiatives. Numerous practical applications highlight companies and organizations investing in projects aimed at restoring degraded ecosystems. These initiatives contribute to Earthmate's holistic approach, emphasizing the interconnected nature of ecological systems. Projects focused on restoring habitats, regenerating biodiversity, and enhancing ecosystem resilience serve as beacons of success. They showcase the positive impact of biodiversity offsetting when aligned with the Earthmate principles, demonstrating that responsible development can coexist with environmental conservation.

Innovations in Carbon Markets and Offset Financing

Successful practices within carbon markets involve the implementation of decentralized models. A case study in the development of decentralized carbon markets showcases platforms facilitating direct engagement between buyers and project developers. This decentralized approach promotes transparency and reduces intermediaries, potentially lowering transaction costs. Earthmate's emphasis on interconnectedness finds resonance in these models, as direct collaboration fosters a more integrated and accountable system, ensuring that the Earthmate principles guide the evolution of carbon markets toward sustainability.

PES

Successful practices within the realm of PES can be observed in community-led conservation projects. For instance, in Costa Rica, the government's PES program encourages landowners to preserve forests and adopt sustainable agricultural practices. Landowners receive payments for the environmental services their land provides, such as carbon sequestration and biodiversity conservation. This initiative not only aligns with the Earthmate principles but also demonstrates the effectiveness of PES in empowering local communities to actively engage in environmental conservation.

Challenges and Obstacles

Individuals

Ordinary People

Despite the positive impact of ordinary people at both the societal and individual levels in adopting the Earthmate principles, several challenges and obstacles can hinder widespread implementation and adoption. Resistance to change, limited access to resources and information, coupled with socio-economic disparities are the important reasons that can impede individuals' ability to adopt sustainable practices effectively. Challenges arise when ordinary people resist or struggle to internalize the Earthmate principles. The biggest hurdles often lie in changing entrenched habits and overcoming resistance to systemic change. For instance, some individuals may struggle to reconcile the delicate balance between individual rights and communal welfare, hindering the principles' acceptance. The principles navigate a delicate balance between individual rights and communal welfare, and not all individuals readily accept the necessary compromises for the greater good. The delicate balance between individual rights and communal welfare poses a challenge, impacting the principles' acceptance. Additionally, some may find it challenging to strike the delicate balance between individual autonomy and collective well-being, fearing the perceived sacrifices required for the greater good.

Other negative aspect is the resistance or reluctance of some individuals to embrace a mindset that prioritizes unity over individualism. The Earthmate philosophy's emphasis on unity despite plurality may face resistance in societies that prioritize individualism or are hesitant to adopt holistic, interconnected approaches. Navigating these obstacles demands a nuanced approach that respects individual rights while emphasizing the broader benefits of a unified, sustainable world. As the philosophy weaves together diverse threads from various philosophical traditions, bridging the gap between theory and practice becomes paramount for its successful adoption. Moreover, negative attitudes or apathy toward the principles can hinder the widespread adoption of sustainable practices. This resistance may be rooted in cultural, economic, or educational factors, reflecting the complexity of influencing diverse populations. Cultural and behavioral barriers may prevent widespread acceptance of the Earthmate principles, particularly in communities where traditional values conflict with environmental conservation efforts. Overcoming ingrained habits and societal norms that contribute to environmental degradation may become a significant obstacle. Overcoming these challenges requires targeted education, awareness campaigns, community engagement, and policy support to empower individuals and communities to understand the practical implications of the principles in their lives and embrace the Earthmate principles fully. Moreover, fostering a culture of inclusivity, collaboration, and respect for diverse perspectives is essential in overcoming resistance to change and driving collective action towards a more sustainable future.

Additionally, disparities in access to resources and information pose challenges, as not all individuals or communities may have equal opportunities to embrace the Earthmate practices. Overcoming these obstacles necessitates a holistic approach that addresses cultural, economic, and educational factors while fostering a sense of shared responsibility for the planet. Additionally, the scalability of successful initiatives may be limited by regional differences and cultural diversity. Bridging these gaps requires targeted efforts in education, communication, and community-building. Initiatives such as community science projects, where citizen scientists contribute to environmental data collection, help address these challenges by fostering a sense of shared responsibility and empowerment. On the practical side, challenges arise in navigating the nuances of UDP in daily life. Some individuals may find it challenging to navigate the nuances of UDP in their daily lives, especially when faced with conflicting interests or deeply ingrained habits. Overcoming these obstacles, that is crucial for creating a more harmonious and interconnected world guided by the principles of UDP, demands raising awareness, providing education, and creating supportive structures. From the micro-level of personal lifestyle choices to the macro-level of community engagement, fostering a Earthmate-driven mindset requires overcoming ingrained patterns of behavior.

Experts and Academia

While the Earthmate principles holds great promise, implementing it faces significant challenges and obstacles. The diverse nature of expertise, as seen in the subcategories like environmental engineers, psychologists, and biodiversity researchers, can lead to conflicting approaches and priorities. Additionally, resistance from entrenched systems and practices poses a formidable obstacle. Overcoming the inertia of established norms requires a concerted effort to shift societal paradigms towards a more interconnected and sustainable mindset. Moreover, the need for global collaboration to address planetary challenges poses a challenge in itself, as geopolitical tensions and varying priorities can impede unified action. Navigating these obstacles necessitates a delicate balance, acknowledging and addressing the complexities inherent in adopting a philosophy that seeks unity despite the plurality of perspectives and interests.

Environmental Writers, Filmmakers and Documentarians

Despite the positive impact of environmental writers, filmmakers, and documentarians, challenges persist in implementing the Earthmate principles through their work. Eco-poets and writers, while contributing to the philosophical discourse, may face the challenge of reaching a broader audience compared to visual mediums. The attention span of audiences, particularly in the age of information overload, poses a hurdle for conveying the depth and nuance inherent in the Earthmate principles. Furthermore, the risk of greenwashing in advocacy-focused filmmaking can undermine the genuine intent of promoting the Earthmate principles, creating skepticism among audiences. Overcoming these challenges necessitates a collaborative effort between environmental communicators, ensuring that their works not only capture attention but also facilitate a meaningful understanding and application of the Earthmate principles in everyday life.

Celebrities, Influencers, Public Figures

Challenges and obstacles arise in implementing the Earthmate principles through celebrity endorsements, influencer partnerships, and online platforms. Eco-friendly travel influencers, sustainable beauty influencers, and zero-waste lifestyle bloggers face the challenge of balancing the aspirational nature of their content with practical, achievable actions that align with the Earthmate principles. Additionally, the risk of greenwashing within these spaces poses a significant obstacle. Discerning the sincerity of influencers' commitment to the Earthmate principles becomes crucial, as superficial gestures can undermine the principles's genuine intent. Navigating these challenges requires a concerted effort to promote authenticity, transparency, and a commitment to long-term, meaningful change within the realms of celebrity influence and online platforms.

Religious and Ethical Leaders

Challenges and obstacles arise in implementing the Earthmate principles within religious and ethical contexts. Green spiritual leaders may face resistance from conservative factions within their religious communities, challenging their efforts to integrate the Earthmate principles into traditional teachings. Interfaith environmental advocates, while fostering collaboration, may encounter difficulties in reconciling divergent beliefs and practices among different faith traditions. Ethical philosophers may face skepticism or pushback from individuals or societies resistant to adopting a more interconnected and ecologically conscious worldview. Bridging these gaps requires delicate navigation, emphasizing the shared values and principles that align with the Earthmate principles while respecting the diverse beliefs within religious and ethical communities.

Environmental Reporters and Journalists

Challenges and obstacles exist within the realm of environmental reporting and journalism, particularly regarding the ethical responsibilities of journalists. Investigative environmental reporters face the challenge of balancing the need to expose issues with the potential negative consequences of sensationalized reporting. The fast-paced nature of digital media can lead to oversimplified narratives that lack the depth necessary to convey the Earthmate principles effectively. Solutions-focused journalists may encounter resistance from those who prioritize sensational stories over nuanced, positive solutions. Overcoming these challenges requires a commitment to ethical reporting, ensuring that the Earthmate principles are upheld in journalistic practices. Journalists need to navigate the delicate balance between informing the public about urgent environmental issues and maintaining a constructive, solutions-oriented approach rooted in the Earthmate principles.

Influential Individuals in Sustainable Business

Challenges and obstacles exist in the journey to implement the Earthmate principles within sustainable business practices. Green entrepreneurs may face resistance from traditional business models that prioritize short-term profits over long-term environmental sustainability. Sustainable tech innovators encounter challenges in balancing the rapid pace of technological advancements with the careful consideration required for Earthmate values. Eco-friendly designers may struggle with market demands that prioritize fast fashion and inexpensive production methods, posing a challenge to Earthmate-inspired ethical practices. Overcoming these obstacles requires a concerted effort to shift industry norms, regulatory frameworks that incentivize sustainability, and increased consumer awareness about the importance of the Earthmate principles in business.

Environmental Activists and Advocates

Challenges and obstacles persist in the journey to implement the Earthmate principles within environmental activism. Conservation campaigners may face resistance from industries prioritizing profit over ecological balance. Affected community advocates may encounter challenges in raising awareness about the interconnectedness of environmental issues with social justice concerns. Policy advocates for environmental justice may struggle against bureaucratic hurdles and resistance to systemic change. The diversity of activism styles can lead to fragmented efforts, hindering a unified adoption of the Earthmate principles across the environmental movement. Youth activists, while bringing fresh perspectives, may face skepticism or dismissiveness from older generations, creating generational divides. Overcoming these obstacles requires a collaborative, inclusive approach, recognizing and respecting the diverse strategies within the environmental activism spectrum while maintaining a shared commitment to the Earthmate values.

Environmental Justice and Climate Justice Advocates

Challenges and obstacles exist within the intersectionality of environmental justice, hindering the complete adoption of the Earthmate principles. Environmental racism activists may face resistance from industries and policymakers unwilling to acknowledge the interconnectedness between systemic discrimination and ecological degradation. Gender and environment advocates may encounter challenges in shifting societal norms that perpetuate gender-based disparities in environmental impacts. Indigenous rights and land defenders may struggle against powerful economic interests that prioritize profit over responsible stewardship. The diverse nature of these challenges requires a comprehensive and inclusive approach, recognizing the interplay between social equity and environmental justice. Overcoming these obstacles necessitates fostering partnerships across diverse communities, encouraging dialogue, and implementing policies that address both social and environmental concerns to truly embody the principles of the Earthmate principles.

Culinary Experts and Food Influencers and Their Recipes

Challenges and obstacles emerge in implementing the Earthmate principles within the culinary world. Sustainable cooking channels may face resistance or skepticism from audiences accustomed to conventional, convenience-oriented culinary practices. Additionally, the availability and affordability of locally sourced, organic ingredients can pose challenges, particularly in regions where such options are limited. The fast-paced nature of social media platforms may also incentivize influencers to prioritize trends over long-term, Earthmate-aligned practices. Overcoming these challenges requires a multifaceted approach, including advocacy for sustainable food policies, promoting local agriculture, and fostering a deeper understanding of the interconnectedness between food choices and the environment. Culinary experts and food influencers can play a pivotal role in addressing these challenges by authentically integrating the Earthmate principles into their content and engaging with their audience to promote positive, sustainable change.

Traditional Knowledge Holders

Challenges and obstacles exist in integrating traditional knowledge into the Earthmate principles, particularly concerning indigenous communities. Indigenous elders and wisdom keepers may face challenges in preserving their oral traditions and passing them on to future generations due to cultural erosion and modernization. Traditional herbalists may encounter issues related to the commercialization of traditional remedies, leading to overharvesting and depletion of natural resources. Indigenous artisans, while contributing to sustainable practices, may face challenges in accessing markets that appreciate and compensate them adequately for their traditional, eco-friendly creations. Overcoming these obstacles requires fostering partnerships that respect and empower indigenous communities, ensuring that their traditional knowledge is acknowledged, preserved, and integrated into broader conservation initiatives. It also calls for addressing systemic issues that disproportionately impact indigenous communities, such as land rights, cultural preservation, and economic empowerment.

Educational and Consultative Institutions

Schools and Environmental Education Programs

Challenges and obstacles exist in implementing the Earthmate principles within educational settings. The variation in the depth and quality of environmental education programs across different education levels poses a challenge to achieving a consistent and widespread understanding of the Earthmate principles. Limited resources and time constraints within the academic curriculum may hinder the integration of comprehensive environmental education. Overcoming these challenges requires prioritizing the integration of the principles of the Earthmate principles into core curricula, ensuring that environmental education is not treated as an add-on but as an essential component of overall learning. Nature-based education programs may face logistical challenges, such as access to natural environments and funding for outdoor learning experiences. Addressing these obstacles involves fostering partnerships between educational institutions, environmental organizations, and local communities to create sustainable and accessible solutions for implementing the principles of the Earthmate principles in education.

Colleges and Universities and Environmental Education Programs

Challenges and obstacles exist in implementing the Earthmate principles within colleges and universities. Varied levels of commitment to sustainability across institutions pose a challenge in achieving a collective adoption of the Earthmate principles. Overcoming this challenge requires fostering a shared commitment to sustainable practices and environmental education at a global scale. Additionally, the availability of resources and funding for environmental programs may vary, impacting the depth and scope of the principles of the Earthmate principles integration. Addressing this requires strategic investments and advocacy for the prioritization of environmental education within higher education budgets. Furthermore, bridging the gap between theoretical knowledge and practical application remains a challenge, emphasizing the importance of experiential learning and hands-on environmental initiatives within academic settings. Encouraging collaboration between institutions and industry partners helps ensure that academic research contributes to real-world solutions and innovations, aligning with Earthmate's call for responsible stewardship.

Informal Online or Offline Environmental Education

Challenges in implementing the Earthmate principles in informal environmental education include the need for quality control and standardization of information. The diverse nature of online platforms may result in inconsistent messaging and varying levels of engagement. Ensuring inclusivity and reaching marginalized communities remains a challenge, requiring targeted efforts to bridge digital divides. The reliability and accuracy of information presented on online platforms may be compromised, demanding increased vigilance to counteract misinformation. Furthermore, sustaining interest and participation in online environmental education programs presents a continuous challenge, necessitating innovative and engaging approaches to capture and retain learners' attention. Environmental health and education initiatives face the challenge of integrating environmental health education into medical and nursing curricula, requiring collaborative efforts between environmental and healthcare sectors to address environmental health issues comprehensively.

Partnerships and Collaborations Targeting Environmental Education

Challenges in implementing the Earthmate principles through partnerships include navigating the potential influence of external agendas on educational content. Ensuring that collaborations maintain a focus on unbiased scientific principles while integrating practical applications requires careful oversight. Struggles may arise in balancing the need for academic independence with the demand for industry-relevant skills. Additionally, ensuring that traditional ecological knowledge is respectfully incorporated into mainstream education without appropriation or distortion poses a challenge. Overcoming these obstacles requires a commitment to transparency, shared values, and a continuous dialogue between all stakeholders involved in collaborative environmental education initiatives. It is essential to navigate the delicate balance between academic rigor, industry applicability, and the preservation of diverse knowledge systems.

Environmental Consultants and Advisory Services

Challenges in implementing the Earthmate principles through environmental consultancy services include the potential conflict between short-term economic interests and long-term environmental goals. Environmental consultants may face obstacles in convincing clients to adopt sustainable practices that might initially appear costlier. Moreover, ensuring that consultants consider the interconnected nature of ecological, social, and economic systems requires constant vigilance. Overcoming these challenges involves promoting awareness and education within the consultancy sector, encouraging a shift towards long-term sustainable strategies. The potential conflict of interest needs addressing to ensure that environmental consultants act as advocates for the Earthmate principles, guiding organizations towards responsible stewardship of the planet.

Think Tanks and Policy Institutes

Challenges in implementing the Earthmate principles through think tanks include the potential for bias or skewed perspectives based on political or economic affiliations. Overcoming these challenges requires a commitment to unbiased research and policy recommendations that prioritize the well-being of the planet. The interdisciplinary nature of the Earthmate principles may also pose challenges in translating complex scientific concepts into practical policies. Effective communication strategies and collaboration with scientists, policymakers, and the public are crucial for overcoming this obstacle. Additionally, ensuring that think tanks prioritize long-term ecological balance over short-term gains is essential for upholding the Earthmate principles. Constant vigilance and adherence to ethical standards are necessary to navigate the intricate landscape of environmental policy development.

R&D Institutions

Challenges in implementing the Earthmate principles through R&D institutions include the potential for research agendas to be influenced by funding sources or political interests. Maintaining scientific integrity, transparency, and a commitment to the long-term well-being of the planet is crucial for overcoming these challenges. The interdisciplinary nature of the Earthmate principles may also pose obstacles in translating complex scientific concepts into practical solutions. Effective communication and collaboration with policymakers, businesses, and the public are essential for addressing this challenge. Additionally, R&D institutions must ensure that their advancements contribute to a circular economy and promote responsible resource consumption. Overcoming these challenges requires a steadfast commitment to the Earthmate principles, ethical conduct, and a focus on holistic and sustainable solutions.

Media and Culture

Media and Communication

Challenges in aligning media outlets with the Earthmate principles include the potential for misinformation, greenwashing, or the oversimplification of complex environmental issues. Mainstream media's reliance on advertising revenue may also hinder unbiased reporting on topics that conflict with corporate interests. Furthermore, ensuring the inclusion of indigenous perspectives and preserving traditional knowledge can be challenging in media narratives. Addressing these obstacles requires a commitment to journalistic integrity, ethical reporting, and the promotion of diverse voices. Media outlets must navigate the complexities of their role, recognizing the influence they wield in shaping public perceptions and contributing to a more sustainable world.

Social Media and Online Activism

Despite its positive impact, implementing the Earthmate principles through social media and online activism faces various challenges. The ease with which misinformation spreads online poses a significant hurdle, potentially undermining the accurate understanding of environmental issues. The risk of performative activism, where individuals engage superficially without substantive contributions, is another challenge. Additionally, the diverse demographic reached through online platforms necessitates tailored communication strategies to ensure inclusivity. Overcoming these challenges requires a concerted effort to promote accurate information, encourage meaningful engagement, and tailor communication to resonate with diverse audiences. Balancing the speed and accessibility of online activism with the depth required for genuine environmental change remains an ongoing consideration.

Religious and Ethical Perspectives

While religious and ethical perspectives offer valuable insights, challenges persist in incorporating these principles into mainstream practices. Divergent interpretations of religious teachings and ethical frameworks can lead to conflicting views on environmental stewardship. Resistance to change, particularly from traditionalist perspectives, may hinder the adoption of more sustainable practices. Additionally, secular societies might face challenges in integrating diverse religious and ethical viewpoints into overarching environmental policies. Recognizing and navigating the complexities arising from diverse beliefs is crucial. Striking a balance between respecting cultural and religious traditions and adapting to contemporary environmental needs poses an ongoing challenge in implementing the Earthmate principles through religious and ethical perspectives.

Art and Literature

Despite the positive impact, challenges persist in fully implementing the Earthmate principles through art and literature. Some artistic expressions may remain niche, limiting their reach and impact. The fashion industry faces obstacles in achieving universal sustainability, with many companies struggling to balance profit margins and ethical practices. Balancing cultural events to be inclusive and representative of diverse perspectives poses a challenge. Additionally, literature and art may sometimes fail to engage certain demographics, limiting the universality of the environmental message. Addressing these challenges requires fostering a culture where sustainable practices are not just a trend but an integral part of artistic and literary expression, ensuring a more profound and widespread impact on the Earthmate principles implementation.

Profit-Oriented Entities

Financial Institutions

Implementing the Earthmate principles in financial institutions faces obstacles rooted in traditional financial models. While impact investing gains traction, some institutions may struggle to balance the dual objectives of financial returns and positive societal and environmental impacts. The adoption of sustainable finance practices requires overcoming resistance within the industry and addressing concerns about reduced profitability. The lack of universal standards for ethical investment and green financing poses a challenge, as definitions and criteria can vary, leading to potential inconsistencies. Overcoming these challenges involves fostering a broader understanding of the interconnected nature of economic, social, and environmental systems within the financial sector.

Industrial Productors

Implementing the Earthmate principles in industrial production faces challenges rooted in traditional practices and economic considerations.

Industries may struggle with transitioning to circular economy initiatives due to the costs and complexities involved. The adoption of sustainable sourcing practices for raw materials poses challenges, as it requires reevaluating supply chains to prevent deforestation and habitat destruction. Overcoming these challenges involves promoting biomimicry – designing products inspired by nature for sustainable solutions. Preferential procurement, where businesses prioritize purchasing from companies with strong environmental practices, can be encouraged to incentivize positive change within industrial sectors.

Mining Industry

Despite positive endeavors, many challenges persist within the mining industry when it comes to fully implementing the Earthmate principles. While some companies champion responsible practices, others may lag, leading to environmental degradation and social conflicts. Differentiating between companies implementing responsible mining practices and those facing sustainability challenges is crucial. Environmental impact assessments must be thorough, addressing potential pitfalls associated with mining activities. The broader challenge involves transforming the industry's mindset and practices, emphasizing long-term sustainability over short-term gains. Promoting widespread adoption of recycled materials and responsible sourcing remains an ongoing obstacle, requiring concerted efforts to shift industry norms and practices.

Agriculture, Agroforestry, Permaculture, Sustainable Agriculture and Forestry Initiatives

Distinctions between traditional, intensive, and industrial agriculture present a challenge, as the shift towards sustainable practices may face resistance from entrenched systems. Encouraging widespread adoption of sustainable agriculture practices requires overcoming obstacles related to awareness, education, and economic incentives. Certification standards, while essential, may face challenges in universal implementation and adherence. Additionally, integrating small and local carbon offset projects, while empowering communities, may encounter hurdles in terms of scalability and standardized measurement. Balancing economic viability with sustainable practices remains a persistent challenge, especially in regions heavily reliant on traditional agricultural methods. Overcoming these challenges necessitates a collaborative effort involving governments, businesses, and communities to transition towards sustainable practices-inspired agricultural and forestry initiatives.

Technological Innovators and Environmental Technology Startups

Despite the positive contributions, technological innovators face challenges and obstacles in aligning with the Earthmate principles. One significant challenge is the entrenched systems within the technology industry that prioritize profit over sustainability. The rapid pace of technological advancements often results in electronic waste, contributing to environmental degradation. Additionally, the environmental impact of manufacturing and resource extraction for technology components poses challenges to achieving a circular economy. Overcoming these obstacles requires a shift in the industry's mindset, promoting sustainable practices, responsible consumption, and the development of technologies with a focus on environmental stewardship. Balancing economic interests with ecological considerations remains a delicate challenge that technology companies must navigate to genuinely adhere to the Earthmate principles.

Tourism Industry, Sustainable Tourism and Ecotourism

Despite the positive intentions, the tourism industry faces challenges and obstacles in aligning with the Earthmate principles. One significant challenge lies in distinguishing between conventional tourism practices and those of ecotourism operators. Greenwashing, where companies present themselves as environmentally friendly without implementing substantial changes, poses a threat to genuine sustainable tourism efforts. Additionally, the sheer scale of the tourism industry, with its potential for overcrowding and habitat disruption, poses a challenge to maintaining delicate ecosystems. Achieving a balance between providing economic opportunities for local communities and preserving the environment remains a complex task. Overcoming these challenges requires a commitment to transparency, education, and collaboration between stakeholders to ensure that sustainable tourism practices genuinely align with the Earthmate principles.

Supply Chains and Consumer Behavior

While progress is evident, challenges persist in fully implementing the Earthmate principles within supply chains. One significant challenge lies in achieving universal supply chain transparency, as not all companies disclose comprehensive information about their sourcing and production practices. The complexity of global supply chains and the prevalence of entrenched practices present obstacles to swift transformation. Balancing economic considerations with sustainability goals remains a challenge, especially for businesses operating within competitive markets. Overcoming these obstacles requires collaborative efforts among businesses, consumers, and regulatory bodies to drive systemic change and align supply chain practices with the principles of the Earthmate principles.

Green Building and Construction

Despite the positive strides, challenges persist in fully implementing the Earthmate principles in the construction industry. The adoption of sustainable practices may face resistance due to entrenched norms within the construction sector, which has historically been resource-intensive and environmentally impactful. Economic considerations often pose challenges, as green building practices may initially incur higher costs. Balancing the need for affordable housing and infrastructure with the long-term benefits of sustainable construction remains a complex challenge. Overcoming these obstacles requires collaborative efforts among stakeholders, including businesses, policymakers, and consumers, to shift the paradigm in the construction industry toward Earthmate-aligned practices.

Packaging and Sustainable Packaging Companies

Despite the positive strides, challenges persist in fully implementing the Earthmate principles in the packaging industry. Resistance to change within established systems, economic considerations, and consumer behavior present obstacles to adopting sustainable packaging solutions. The use of conventional packaging materials, such as single-use plastics, is deeply ingrained in existing supply chains, and transitioning to sustainable alternatives may face initial resistance. Economic factors, including the perceived cost of sustainable packaging materials, can pose challenges for businesses, particularly smaller ones with limited resources. Additionally, consumer awareness and preferences play a pivotal role in driving the demand for sustainable packaging, and efforts to educate and shift consumer behavior are ongoing. Overcoming these obstacles requires collaborative efforts among businesses, consumers, and policymakers to create an environment that supports the adoption of Earthmate-aligned packaging practices.

Waste Management Industry

Despite the positive contributions, challenges persist in fully implementing the Earthmate principles within the waste management industry. The historical reliance on traditional waste disposal methods, coupled with existing infrastructure, poses resistance to adopting more sustainable practices. Economic considerations, such as the cost of implementing advanced recycling technologies, can be a significant obstacle for some companies. Additionally, public awareness and education play a crucial role in overcoming resistance to change, as communities may be resistant to new waste management practices. Collaborative efforts among governments, businesses, and communities are essential to navigating these challenges and fostering an environment conducive to the widespread adoption of the Earthmate principles within the waste management sector.

Renewable Energy Industry

While the renewable energy industry contributes positively to the Earthmate principles, challenges persist in its widespread adoption. Economic considerations, such as the initial costs of renewable energy infrastructure, pose barriers to entry for some regions and communities. Additionally, the intermittent nature of certain renewable sources, like solar and wind, presents challenges in maintaining a consistent energy supply. Overcoming these obstacles requires a nuanced understanding of the interconnected systems at play. The industry must navigate political, technological, and economic landscapes to ensure a delicate balance between economic prosperity, social well-being, and environmental protection. Collaborative efforts among governments, businesses, and communities are essential to address these challenges and fully embrace the Earthmate principles within the renewable energy sector.

PPPs

While PPPs offer significant benefits, challenges and obstacles exist in their journey to implement the Earthmate principles. Balancing profit-oriented objectives with environmental and social considerations requires careful navigation. Potential conflicts of interest, differing priorities, and varying levels of commitment among partners may hinder progress. The delicate balance needed for societies to thrive, as emphasized by the Earthmate principles, is particularly challenging in PPPs where diverse stakeholders with distinct motivations are involved. The potential for power imbalances between private and public entities further complicates the collaborative landscape. Navigating these challenges demands a nuanced understanding of the interplay between individual rights and communal welfare. By acknowledging and addressing these obstacles,

PPPs can evolve into powerful instruments for sustainable development, aligning with Earthmate's overarching vision for a unified and interconnected world.

Non-profit Entities

Environmental Justice and Climate Justice Associations

Despite their positive impact, Environmental Justice and Climate Justice Associations face challenges and obstacles in fully implementing the Earthmate principles. These challenges may include resistance from entrenched systems and structures that perpetuate environmental injustices. Limited resources, both financial and institutional, can hinder the effectiveness of these organizations in addressing complex issues. Additionally, navigating legal frameworks and influencing policies requires persistent efforts, and the potential for power imbalances in decision-making processes can be a barrier. Recognizing these challenges is essential for the Earthmate principles to adapt and evolve, ensuring that justice associations can effectively contribute to a world where unity prevails amid the plurality of existence.

Wildlife Conservation Associations

Despite their positive contributions, Wildlife Conservation Associations encounter challenges and obstacles in fully implementing the Earthmate principles. Challenges may include insufficient funding, limited resources, and the need for policy advocacy to influence legal frameworks positively. Engaging local communities in conservation efforts, while beneficial, can present obstacles such as cultural sensitivities, power imbalances, or resistance to change. Navigating these challenges is essential for the Earthmate principles to adapt and evolve, ensuring that wildlife conservation efforts effectively contribute to a world where unity prevails amid the plurality of existence. The recognition of potential pitfalls, as the Earthmate principles acknowledges, underscores the need for nuanced and collaborative approaches in wildlife conservation.

Climate Change Mitigation and Adaptation Associations

Despite their positive contributions, climate change mitigation and adaptation associations face challenges and obstacles in fully implementing the Earthmate principles. Distinguishing between organizations focused on mitigation and adaptation highlights potential conflicts in priorities and resource allocation. Innovative approaches to climate resilience and adaptation may encounter resistance or face barriers to widespread adoption. Policy advocacy and legal support are crucial elements in influencing environmental legislation and regulations, but the complexities of political and legal processes can present obstacles. Navigating these challenges is essential for the Earthmate principles to adapt and evolve, ensuring that climate change efforts effectively contribute to a world where unity prevails amid the plurality of existence. The acknowledgment of potential pitfalls underscores the need for nuanced and collaborative approaches in addressing climate change.

Ocean Protection Associations

Despite the positive contributions of Ocean Protection Associations, challenges and obstacles persist in fully implementing the Earthmate principles. The global scale of issues such as overfishing and plastic pollution requires comprehensive international cooperation, making it challenging to establish unified regulations. Additionally, the rapid degradation of marine ecosystems demands swift and decisive action, yet bureaucratic hurdles and conflicting economic interests often impede progress. Earthmate's emphasis on the delicate balance between economic prosperity, social well-being, and environmental protection underscores the need to overcome these obstacles through collective efforts and mindful decision-making. Non-profit organizations engaged in policy advocacy and legal support play a crucial role in influencing environmental legislation, regulations, and legal frameworks to address these challenges effectively.

Community-Based Conservation Initiatives

While Community-Based Conservation Initiatives demonstrate positive impacts, they also face challenges and obstacles in fully implementing the Earthmate principles. One significant challenge lies in ensuring that conservation practices are sustainable and beneficial to both the environment and local communities. Balancing economic interests, cultural sensitivities, and the need for biodiversity conservation can be intricate. Moreover, bureaucratic hurdles and conflicting policies may hinder the swift and effective implementation of these initiatives. Non-profit organizations engaged in policy advocacy and legal support play a crucial role in influencing environmental legislation, regulations, and legal frameworks to address these challenges effectively and ensure that the Earthmate principles are upheld.

Green Philanthropy and Impact Investing

While green philanthropy and impact investing contribute positively, challenges and obstacles exist in fully implementing the Earthmate principles through these financial models. One significant challenge lies in ensuring that financial resources are directed towards initiatives that genuinely adhere to Earthmate's holistic and interconnected approach. There may be conflicts between solely profit-driven motives and the environmental objectives inherent in the Earthmate principles. Additionally, navigating regulatory frameworks and ensuring transparency in impact measurement can pose hurdles. Overcoming these challenges requires a concerted effort from organizations engaged in green philanthropy and impact investing, emphasizing the need for ethical, sustainable, and Earthmate-aligned investment practices.

Legal Advocacy and Environmental Law NGOs

Despite the positive contributions, challenges and obstacles exist in fully implementing the Earthmate principles through legal advocacy and Environmental Law NGOs. One significant challenge lies in navigating complex legal systems and overcoming barriers to litigation against powerful entities responsible for environmental harm. Resource constraints and the need for long-term, sustained efforts can impede the effectiveness of legal actions. Additionally, there may be conflicts between economic interests and the environmental objectives advocated by these organizations. Overcoming these challenges requires continuous efforts to strengthen legal frameworks, promote public awareness, and foster collaborations between diverse stakeholders. Legal advocacy, grounded in the Earthmate principles, should navigate these obstacles to ensure a balanced and interconnected approach to environmental protection.

Crisis Response and Disaster Relief NGOs

Despite their positive contributions, Crisis Response and Disaster Relief NGOs face challenges and obstacles in fully implementing the Earthmate principles. One significant challenge lies in the increasing frequency and intensity of climate-related disasters, demanding innovative and adaptive responses. Coordination among various relief organizations, government agencies, and communities remains a complex task, highlighting the need for a comprehensive and inclusive strategy. Additionally, issues related to equity and social justice may arise during the distribution of relief, necessitating a nuanced approach to ensure fair and equitable support for all affected populations. Navigating these challenges requires continuous efforts to enhance the effectiveness of crisis response mechanisms, promote public awareness, and foster collaborations between diverse stakeholders. Crisis Response and Disaster Relief NGOs must work towards aligning their actions with the Earthmate principles, emphasizing the interconnectedness of environmental well-being and human welfare. Overcoming these obstacles is crucial to ensuring that crisis response efforts contribute to Earthmate's vision of a sustainable and resilient future for both humanity and the planet.

Governmental Entities

Legislative Bodies

Despite their positive contributions, legislative bodies face challenges and obstacles in fully implementing the Earthmate principles. The complexities of political dynamics and conflicting interests may hinder the development of comprehensive and effective environmental policies. Striking a balance between individual rights and communal welfare, as emphasized by the Earthmate principles, can be challenging within the legislative process. Overcoming these obstacles requires a nuanced understanding of the interconnected nature of environmental issues and a commitment to overcoming political and economic barriers. Emphasizing the importance of legislative bodies in promoting sustainable policies and addressing environmental challenges encourages ongoing efforts to align legislative actions with the Earthmate principles.

Legal and Regulatory Frameworks

Despite their positive contributions, legal and regulatory frameworks face challenges and obstacles in fully implementing the Earthmate principles. One major challenge is the need for stringent enforcement, ensuring that regulations are not only comprehensive but are also actively applied. Governments and regulatory bodies must balance the need for strict regulations with practical enforcement mechanisms. Additionally, navigating political, economic, and cultural complexities poses obstacles to the effectiveness of legal frameworks. Encouraging governments to assess and update regulations regularly is crucial for overcoming these challenges and ensuring that legal frameworks remain adaptable and responsive to evolving environmental issues. The recognition of both successes and challenges within legal and regulatory frameworks emphasizes the ongoing efforts needed to align governance structures with the principles of the Earthmate principles.

Law Enforcement Entities

Despite their positive contributions, law enforcement entities face challenges and obstacles in fully implementing the Earthmate principles. The most rudimentary and vital challenges within law enforcement entities involve combatting corruption, ensuring transparency, eliminating bias, and mitigating individual and party interests. These challenges should be addressed with a commitment to upholding the principles of justice, fairness, and public trust. A rigorous focus on eradicating corruption within law enforcement is paramount to maintaining environmental protection. Transparency, as a foundational principle, is essential for fostering public confidence and accountability. Simultaneously, efforts to eliminate bias and mitigate personal and party interests are crucial to ensuring that law enforcement functions impartially and in the best interests of the community. This multifaceted approach not only enhances the credibility of law enforcement entities but also contributes to the establishment of a just and equitable society in fully implementing the Earthmate principles. Governments involved in infrastructure development must navigate the delicate balance between progress and eco-friendly practices, especially concerning habitat disruption. Enforcing strict environmental regulations to prevent illegal activities requires continuous efforts, and effective policies, including penalties for biodiversity-related violations, need to be established and maintained. Engaging local communities in environmental protection efforts demands overcoming cultural and socio-economic barriers. Encouraging governments to implement incentives for green technologies and preferential procurement practices is essential, but challenges may arise in establishing and enforcing these policies effectively. The recognition of both successes and challenges within law enforcement entities emphasizes the ongoing efforts needed to align law enforcement practices with the Earthmate principles.

Judiciary

Despite the positive role of the judiciary, challenges exist in accessing legal avenues for citizens and organizations seeking justice in environmental matters. Legal procedures can be complex and costly, hindering accessibility for many. Additionally, inconsistent enforcement of environmental laws poses a challenge to the effective implementation of the Earthmate principles. In some cases, weak enforcement mechanisms may undermine the impact of well-intentioned legislation, highlighting the need for ongoing efforts to strengthen legal frameworks and ensure their practical effectiveness in addressing environmental challenges.

Governmental Organizations for Wildlife Conservation

While governmental organizations play a crucial role, challenges exist in implementing the Earthmate principles in wildlife conservation. Limited resources, both financial and human, often hinder the effectiveness of conservation efforts. Additionally, political pressures and conflicting priorities may compromise the long-term vision of the Earthmate principles. Addressing these challenges requires a commitment to sustainable resource allocation, international cooperation, and a holistic understanding of the interconnected systems that govern wildlife conservation. Overcoming these obstacles is essential to ensuring the preservation of biodiversity and fostering a harmonious coexistence between humanity and the natural world.

Government Water Resource Management Entities

Despite positive efforts, challenges persist in implementing the Earthmate principles in water resource management. One major obstacle is the difficulty of ensuring equitable water use, particularly in regions with competing demands and scarce resources. Political and economic pressures can compromise the implementation of integrated plans, hindering the achievement of a harmonious balance between ecological and human needs. Overcoming these challenges requires a concerted effort to foster understanding, collaboration, and the development of innovative solutions that align with the Earthmate principles.

Governmental Institutions for Meteorology, Climate Science, and Weather Forecasting

Despite positive contributions, challenges persist in implementing the Earthmate principles within meteorological institutions. One significant challenge is the unpredictability and complexity of Earth's climate system. Developing accurate climate models and forecasting systems requires ongoing advancements in technology and interdisciplinary collaboration. Additionally, political and economic pressures may impact the autonomy and funding of these institutions, hindering their ability to address climate change comprehensively. Overcoming these challenges necessitates a commitment to fostering global cooperation, overcoming political barriers, and advancing scientific research to align with the Earthmate principles.

Environmental Government Agencies

Despite positive efforts, environmental government agencies encounter challenges in fully implementing the Earthmate principles. One major obstacle is the trade-off between economic development and environmental conservation. Agencies may struggle to strike a balance, leading to compromises that undermine the Earthmate principles' emphasis on responsible stewardship. Additionally, bureaucratic inefficiencies, limited resources, and inconsistent enforcement of regulations hinder agencies' ability to address environmental issues comprehensively. Overcoming these challenges requires a concerted effort to prioritize transparency, public engagement, and the integration of the Earthmate principles into decision-making processes. It involves fostering a culture of environmental awareness and responsibility within governmental structures.

Urban Planning, Green Infrastructure and Sustainable Development

Despite positive strides, challenges and obstacles persist in implementing the Earthmate principles in urban planning and sustainable development. One major challenge is the conflict between economic interests and environmental conservation. Urbanization often comes with the pressure to prioritize economic growth, leading to unsustainable practices that compromise the Earthmate principles' emphasis on responsible stewardship. Additionally, urban sprawl, inadequate infrastructure, and lack of public awareness pose obstacles to creating sustainable urban spaces. Overcoming these challenges requires a holistic approach, involving collaborative efforts from governments, urban planners, and the community. Implementing the Earthmate principles in urban development demands a paradigm shift, where economic progress aligns with ecological sustainability for the benefit of present and future generations.

Health and Environment

Healthcare Industry

Despite progress, challenges and obstacles hinder the healthcare industry's complete adoption of the Earthmate principles. Barriers include the industry's heavy reliance on single-use plastics and the environmental impact of pharmaceutical waste. Overcoming these challenges requires a paradigm shift towards circular economy principles and sustainable drug disposal practices. Additionally, the interconnectedness of human, animal, and environmental health, while recognized as essential, poses logistical challenges in implementing collaborative "One Health" approaches. Balancing healthcare advancements with environmental conservation remains an ongoing challenge, emphasizing the need for continued efforts to integrate the Earthmate principles into the core practices of the healthcare industry.

Medical Infrastructure and Health Systems

Despite positive efforts, challenges and obstacles persist in fully implementing the Earthmate principles within medical infrastructure and health systems. One challenge involves addressing health disparities, especially in marginalized communities disproportionately affected by environmental hazards. Advocacy for health policies that recognize and mitigate these disparities requires overcoming systemic barriers and biases. Additionally, the integration of climate resilience strategies may face resistance due to resource constraints and competing priorities within health systems. Achieving equitable access to green spaces in urban planning presents challenges related to land use, community engagement, and resource allocation. Overcoming these obstacles requires continued advocacy, collaboration, and a commitment to balancing the interconnected needs of human health and environmental sustainability.

Environmental Psychology

Despite the positive contributions, challenges and obstacles exist in fully implementing the Earthmate principles within the realm of environmental psychology. One challenge involves overcoming resistance to nature-based interventions in traditional healthcare practices, where the value of green spaces may be underestimated. Additionally, achieving collaboration between environmental psychologists and urban planners may face barriers related to differing priorities, methodologies, and language. Overcoming these challenges requires interdisciplinary efforts, effective communication, and a shared commitment to fostering environments that prioritize both human well-being and environmental sustainability.

Community-Based Health Initiatives

Despite the positive contributions, challenges and obstacles exist in implementing the Earthmate principles through community-based health initiatives. One challenge involves overcoming potential resistance or lack of awareness within communities about the interconnected nature of environmental and public health. Additionally, ensuring the long-term sustainability of these initiatives may face hurdles related to funding, resource constraints, and changing community dynamics. Addressing these challenges requires a comprehensive approach that includes community education, stakeholder collaboration, and ongoing support to empower communities to overcome obstacles and drive positive change.

Global Health and Environmental International Organizations

While global health and environmental organizations contribute positively, challenges and obstacles exist in implementing the Earthmate principles on a global scale. One challenge involves navigating geopolitical complexities and differing priorities among nations, hindering the seamless adoption of unified the Earthmate principles. Additionally, resource constraints, political resistance, and the lack of a universally accepted framework can impede progress. The interdisciplinary nature of the Earthmate principles demands overcoming silos within organizations and aligning diverse perspectives. Striking a balance between individual autonomy and collective well-being on a global scale poses a significant challenge, requiring ongoing efforts to foster a shared understanding and commitment to the Earthmate principles among diverse nations and stakeholders.

International Organizations

United Nations and UN Specialized Agencies

Despite the positive strides, challenges persist in implementing the Earthmate principles through UN agencies. The fragmentation of efforts among various specialized agencies and the lack of a unified approach can lead to overlapping mandates and inefficiencies. In this field,

Maurice Strong emphasized the need for a new integrative management approach to address complex environmental issues that transcend traditional hierarchical structures, advocating for greater horizontal and transsectoral communication, reflecting the Stockholm leadership's strategy and suggesting future directions for social organization with significant implications for international law and relations (Caldwell and Weiland 1996). Additionally, political considerations often hinder the adoption of bold and transformative policies, reflecting the struggle to balance national interests with global environmental imperatives. Bridging the gap between developed and developing nations in resource allocation and technology transfer remains a significant hurdle. The Earthmate principles' call for collective responsibility faces resistance in instances where short-term national gains conflict with long-term global benefits. Addressing these challenges requires a concerted effort to reform structures, enhance collaboration, and foster a shared sense of responsibility for Earth's well-being.

Global Specialized Conservation and Biodiversity Organizations

While these organizations contribute significantly to the Earthmate principles, challenges persist in their implementation. The lack of universal adherence to international agreements and the insufficient enforcement mechanisms hinder the effectiveness of conservation efforts. Financial constraints and disparities in resource allocation among member states pose challenges to implementing the Earthmate principles universally. Balancing conservation goals with economic development aspirations of nations remains a complex task, underscoring the need for a nuanced understanding of the interconnectedness between ecological sustainability and socio-economic factors. Overcoming these obstacles requires ongoing efforts to enhance global cooperation, promote equity, and address the root causes of biodiversity loss.

Non-environmental Organization and Environmental Issues

Despite their positive contributions, non-environmental organizations face challenges in fully adopting the Earthmate principles. One challenge is the potential conflict between economic interests and environmental sustainability. Organizations rooted in traditional economic models may find it challenging to align with Earthmate's call for a delicate balance between economic prosperity and environmental protection. Additionally, navigating diverse stakeholder interests and conflicting priorities poses obstacles. Striking a balance between social, economic, and environmental considerations requires overcoming entrenched structures and practices. These challenges highlight the need for ongoing dialogue, education, and collaboration to facilitate a smoother integration of the Earthmate principles across diverse organizational contexts.

International Environmental Legal Instruments

While International Environmental Binding Legal Instruments and Potentially Binding or Non-Binding Instruments contribute positively, challenges and obstacles persist in fully implementing the Earthmate principles. One challenge is the varying levels of commitment among nations, reflecting the delicate balance needed between individual rights and communal welfare. Implementing principles like common but differentiated responsibilities encounters resistance from nations with differing economic and developmental priorities. For example, the nuclear test cases demonstrate that not all states take their international responsibilities seriously, as reflected in Principle 21 of the Stockholm Declaration, yet the declaration has nonetheless been an effective catalyst for developing a global environmental ethic (Caldwell and Weiland 1996). The need for nuanced understanding, as emphasized in the Earthmate principles, is challenged by geopolitical complexities and competing national interests. Additionally, the lack of effective enforcement mechanisms poses an obstacle, requiring continuous efforts to strengthen global governance for environmental issues. Overcoming these challenges requires persistent diplomacy, education, and the cultivation of a shared Earthmate vision among nations.

Global Environmental Funds and Financing Mechanisms

Despite the positive contributions, challenges and obstacles exist in implementing the Earthmate principles through global environmental funds. One challenge is the equitable distribution of funds, ensuring that marginalized communities within countries receive adequate support. The bureaucratic processes and conditionalities associated with accessing funds may hinder the timely implementation of projects, requiring streamlined procedures aligned with Earthmate's call for efficient and cooperative action. Additionally, the complexity of project evaluation and the need for stringent accountability may pose challenges in striking the right balance between oversight and facilitating impactful initiatives. Overcoming these obstacles requires continuous refinement of funding mechanisms, fostering transparency, and enhancing inclusivity to truly embody the Earthmate principles.

Humanitarian and Environmental Aid Organizations

One challenge faced by humanitarian and environmental aid organizations is balancing immediate needs with long-term sustainability. While organizations focus on providing urgent relief during crises, there is a risk of neglecting the broader environmental context and the underlying factors contributing to disasters. Striking the right balance requires a nuanced understanding of the Earthmate principles, emphasizing both the immediate well-being of communities and the long-term resilience of ecosystems. Additionally, coordinating efforts across diverse sectors, from healthcare to environmental conservation, poses logistical challenges. Overcoming these obstacles necessitates a comprehensive approach that integrates the Earthmate principles into the core of humanitarian and environmental aid strategies, ensuring a harmonious balance between addressing immediate crises and fostering sustainable practices for the future.

International Research and Scientific Collaboration

Despite the positive contributions of international research and scientific collaboration, challenges exist in implementing the Earthmate principles, particularly concerning technology transfer and capacity building. While advocating for organizations to facilitate the transfer of environmentally sustainable technologies to developing countries, it is crucial to address the potential disparities in technological capabilities. Ensuring that these technologies are effectively utilized and adapted to local contexts requires comprehensive capacity-building initiatives. Overcoming these challenges involves integrating the Earthmate principles into strategies for technology transfer, emphasizing inclusivity, and recognizing the diversity of needs and capabilities across nations. Balancing technological advancements with the principles of responsible stewardship and unity despite plurality remains a complex but necessary task.

Special Environment-Oriented Entities

Standardization and Certification

While advocating for independent audits to verify emissions reductions is a positive step, challenges may arise in ensuring universal adherence to such audits. The potential for variability in auditing practices and the emergence of unregulated offset projects can hinder the effectiveness of Earthmate-aligned initiatives. Additionally, reliance on innovative technologies, such as satellite monitoring and artificial intelligence, may face challenges related to affordability, accessibility, and global implementation. Balancing the adoption of cutting-edge technologies with inclusivity and fairness is essential to overcoming obstacles in implementing the Earthmate principles within standardization and certification processes.

Carbon Offset and Environmental Credit Providers

Despite the positive contributions of carbon offset providers, challenges exist in fully aligning with the Earthmate principles. The promotion of LCAs for carbon offset projects becomes essential in evaluating their overall environmental impact comprehensively. Incorporating emissions beyond carbon dioxide, such as methane and nitrous oxide, presents a challenge due to the complexities involved in quantifying and addressing these diverse emissions. Additionally, while carbon offsetting serves as a transition strategy for businesses, ensuring concurrent efforts to reduce internal emissions and improve sustainability practices remains a hurdle. The integration of carbon offsetting into broader corporate sustainability strategies becomes imperative, demanding careful planning and implementation to guarantee alignment with Earthmate's holistic approach to environmental protection.

Biodiversity Offsetting

Despite the positive aspects, challenges and obstacles exist in implementing biodiversity offsetting aligned with the Earthmate principles. One significant challenge is the quantification and equivalency of biodiversity. Unlike carbon offsetting, where emissions are measured in a standardized unit (tons of CO_2), assessing the loss or gain of biodiversity involves complex ecological factors. Determining the equivalency between the impact of a development project and the benefits of an offset program demands nuanced understanding and precise metrics, posing a challenge to the effective implementation of biodiversity offsetting within Earthmate's holistic framework.

Innovations in Carbon Markets and Offset Financing

Despite the positive strides, challenges exist in implementing the Earthmate principles within carbon markets. One obstacle is the need for diverse financing models beyond traditional offset credits. While blockchain and decentralized markets offer advancements, there is a necessity to explore and establish innovative financing mechanisms. This involves identifying financial instruments that support carbon offset projects effectively. The challenge lies in navigating the financial landscape to encourage continuous innovation and secure funding for projects aligned with Earthmate's principles of holistic environmental stewardship.

PES

Despite the positive aspects, challenges persist in implementing PES programs in a way that truly aligns with the Earthmate principles. One prominent challenge is the need to strike a delicate balance between economic interests and ecological well-being. Determining fair and equitable compensation for ecosystem services while ensuring the sustainability of such programs demands careful consideration. Additionally, addressing potential disparities in the distribution of benefits and navigating the complexities of evaluating the multifaceted contributions of ecosystems pose obstacles to the effective implementation of PES in harmony with the Earthmate principles.

Conclusion

In conclusion, the adoption and advancement of Earthmate principles necessitate the collective efforts of individuals, communities, and organizations across various sectors. Ordinary individuals serve as powerful agents of change by embodying Earthmate philosophy in their daily lives and promoting awareness of the interconnectedness between humanity and nature. Educational institutions, media outlets, and cultural influencers play pivotal roles in disseminating knowledge, fostering understanding, and shaping perceptions towards Earthmate principles, despite challenges such as misinformation and industry influence. Profit-oriented entities are increasingly embracing sustainability practices, yet face obstacles such as reluctance to fully adopt Earthmate principles and prioritize short-term gains over long-term ecological sustainability. Non-profit organizations exemplify Earthmate's emphasis on responsible stewardship through positive contributions to environmental conservation and community empowerment, despite challenges like funding constraints and legal obstacles. Governmental entities play crucial roles in adopting Earthmate principles through policies, regulations, and urban planning initiatives, although political pressures and limited resources pose challenges. Additionally, across sectors including healthcare, environmental psychology, and global health organizations, stakeholders strive to prioritize environmental sustainability and collective well-being, despite challenges like healthcare waste and psychological barriers. Ultimately, the collective commitment to collaboration, innovation, and responsible decision-making underscores the resilience of the Earthmate philosophy, paving the way for a sustainable and interconnected future.

Furthermore, successful practices and initiatives across media and culture, profit-oriented entities, non-profit organizations, governmental bodies, health and environmental sectors, international organizations, and special environment-oriented entities collectively demonstrate the effectiveness of adopting Earthmate principles in fostering sustainability and interconnectedness. These initiatives highlight the transformative potential of integrating environmental stewardship into various aspects of human endeavors, from education and policy development to media representation and corporate practices. By recognizing and amplifying these success stories, individuals, communities, and organizations worldwide can draw inspiration and cultivate a shared commitment to Earthmate principles, ultimately paving the way for a more resilient, equitable, and harmonious world for present and future generations. Through collaborative efforts and a holistic approach, the vision of Earthmate can be realized, promoting the well-being of both humanity and the planet.

Finally, while the Earthmate principles provide a visionary path towards a sustainable and interconnected world, a myriad of challenges and obstacles must be addressed to fully realize their potential across diverse sectors and communities. Individuals face resistance to change, limited resources, and cultural barriers, hindering the adoption of sustainable practices in daily life. Experts and academia encounter conflicting priorities and geopolitical tensions, impeding unified action towards sustainability. Environmental communicators and influencers face greenwashing and authenticity issues, while religious and ethical leaders navigate diverse beliefs within communities. Sustainable businesses and activists confront resistance from traditional models and systemic injustices. Across sectors like education, media, profit-oriented entities, non-profit organizations, government bodies, health, and the environment, challenges persist, from funding constraints to political dynamics. Overcoming these obstacles demands a multifaceted approach, prioritizing education, collaboration, inclusivity, and policy support to empower individuals, communities, and institutions to embrace Earthmate principles fully. Only through concerted efforts can we pave the way towards a harmonious and sustainable future for all, inspired by the transformative potential of unity and responsible stewardship advocated by the Earthmate principles.

References

Caldwell, L. K., & Weiland, P. S. (1996). *International environmental policy: from the twentieth to the twenty-first century.* Duke University Press.

Gidlöf-Gunnarsson, A., & Öhrström, E. (2007). Noise and well-being in urban residential environments: The potential role of perceived availability to nearby green areas. *Landscape and urban planning*, 83(2-3), 115-126.

Guite, H. F., Clark, C., & Ackrill, G. (2006). The impact of the physical and urban environment on mental well-being. *Public health*, 120(12), 1117-1126.

Guo et al., W. (2021). Effect of ambient air quality on subjective well-being among Chinese working adults. *Journal of Cleaner Production*, 296, 126509.

Jabbar, M., Yusoff, M. M., & Shafie, A. (2022). Assessing the role of urban green spaces for human well-being: A systematic review. *GeoJournal*, 1–19.

Lal et al., R. M. (2020). Connecting air quality with emotional well-being and neighborhood infrastructure in a US city. *Environmental Health Insights*, 14, 1178630220915488.

Ma et al. (2019). Effects of urban green spaces on residents' well-being. *Environment, Development and Sustainability*, 21, 2793–2809.

Maalouf, A., & Mavropoulos, A. (2023). Re-assessing global municipal solid waste generation. *Waste Management & Research*, 41(4), 936-947.

Reyes-Riveros et al., R. (2021). Linking public urban green spaces and human well-being: A systematic review. *Urban forestry & urban greening*, 61, 127105.

Smil, V. (1994). How many people can the earth feed? *Population and Development Review*, 255–292.

UNEP. (2023). *State of Finance for Nature: The Big Nature Turnaround –Repurposing $7 trillion to combat nature loss.* Nairobi.

Yuan, L., Shin, K., & Managi, S. (2018). Subjective well-being and environmental quality: the impact of air pollution and green coverage in China. *Ecological economics*, 153, 124-138.

6

The Future of Earthmate

The Potential Impact of the Earthmate Principles

In the previous season, it was articulated that the Earthmate principles encompass three fundamental aspects: Earthmate Philosophy, Earthmate Science, and Earthmate Ethics. Within this framework, we delve into the possible ramifications of embracing the Earthmate principles across these three distinct domains.

The potential impact of widespread adoption of the Earthmate Philosophy is profound and far-reaching, offering a holistic framework for fostering harmony, sustainability, and well-being on a global scale. By embracing the principles of UDP and recognizing the interconnectedness of all beings and systems, societies can cultivate a deeper sense of empathy, cooperation, and stewardship towards the planet and its inhabitants. One significant impact would be the promotion of a more inclusive and collaborative approach to addressing environmental and social challenges. By emphasizing the interconnectedness of all living things, the Earthmate Philosophy encourages individuals and communities to work together across differences and boundaries to achieve common goals. This collaborative ethos can lead to the development of innovative solutions, the sharing of resources and knowledge, and the building of resilient communities capable of adapting to change. Furthermore, widespread adoption of the Earthmate Philosophy could foster a greater sense of responsibility and accountability towards the planet. By recognizing the intrinsic value of nature and the need for responsible stewardship, societies can prioritize environmental conservation and sustainable development. This could lead to the protection of ecosystems, the preservation of biodiversity, and the promotion of regenerative practices that restore and replenish natural resources. Moreover, the Earthmate Philosophy has the potential to inspire a cultural shift towards values of equity, justice, and compassion. By emphasizing principles such as respect for nature, interdependence, and sustainability, societies can address social injustices and inequities that exacerbate environmental degradation. This could lead to more equitable distribution of resources, greater social cohesion, and a sense of solidarity among diverse communities. Overall, the widespread adoption of the Earthmate Philosophy has the potential to catalyze a transformation in how societies relate to the planet and each other. By fostering a deeper understanding of our interconnectedness and shared responsibility, this philosophy offers a pathway towards a more sustainable, equitable, and harmonious world for present and future generations.

The potential impact of widespread adoption of Earthmate Science is vast and transformative, offering a comprehensive approach to addressing the complex environmental challenges facing our planet. By embracing principles rooted in scientific rigor and interdisciplinary collaboration, societies can unlock innovative solutions, foster sustainable development, and safeguard the health of ecosystems and communities. One significant impact would be the development of more effective and holistic approaches to environmental management and conservation. Earthmate Science, grounded in Complex Systems and Systems Thinking, recognizes the interconnected nature of Earth's systems. By integrating knowledge from diverse scientific disciplines, such as ecology, climatology, geology, and sociology, societies can gain a deeper understanding of the complex interactions driving environmental change. This holistic perspective enables the development of more effective strategies for mitigating environmental degradation, conserving biodiversity, and promoting ecosystem resilience. Furthermore, widespread adoption of Earthmate Science could drive technological innovation and research breakthroughs in areas critical to environmental sustainability. By prioritizing research and development in fields such as renewable energy, clean technologies, and sustainable agriculture, societies can accelerate the transition towards a low-carbon and resource-efficient economy. This could lead to significant reductions in GHG emissions, pollution, and resource depletion, while also creating new economic opportunities and industries. Moreover, Earthmate Science has the potential to foster greater international cooperation and collaboration on environmental issues. By promoting knowledge sharing, capacity building, and scientific diplomacy, societies can work together to address common environmental challenges, such as climate change, deforestation, and ocean pollution. This collaborative approach can lead to the development of global frameworks and agreements for environmental protection, ensuring that collective action is taken to safeguard the health of the planet for future generations. Overall, the widespread adoption of Earthmate Science has the potential to catalyze a paradigm shift in how societies approach environmental stewardship. By harnessing the power of science and innovation, societies can build a more sustainable and resilient future, where the needs of people and the planet are balanced and harmonized for the benefit of all.

The potential impact of widespread adoption of the Earthmate Ethics is far-reaching and profound, as it offers a transformative framework for guiding individual and collective behavior towards a more sustainable and harmonious relationship with the planet. By embracing the principles of respect for nature, interdependence, responsibility, sustainability, equity, and justice, societies can address pressing environmental challenges and foster long-term well-being for both people and the planet. One significant impact would be a shift towards more conscientious and sustainable consumption patterns. By prioritizing sustainability and recognizing the finite nature of resources, individuals and communities would make choices that minimize waste, conserve resources, and reduce environmental degradation. This shift could lead to significant reductions in pollution, habitat destruction, and biodiversity loss, thereby preserving ecosystems and safeguarding the health of the planet. Furthermore, widespread adoption of Earthmate Ethics could promote social cohesion and collaboration on a global scale. By emphasizing shared goals and values, such as the well-being of present and future generations, communities would be motivated to work together across borders and differences to address common environmental challenges. This collaboration could lead to the development of innovative solutions, the sharing of knowledge and resources, and the establishment of more effective governance structures for environmental stewardship. Moreover, Earthmate Ethics could foster greater equity and justice in the distribution of environmental benefits and burdens. By prioritizing fairness and inclusivity, the adoption of these ethical principles could help address environmental injustices, such as disproportionate exposure to pollution and the unequal distribution of natural resources. This could lead to more equitable outcomes for marginalized communities and ensure that the benefits of environmental protection are shared by all. Overall, the widespread adoption of Earthmate Ethics has the potential to catalyze a fundamental shift towards a more sustainable, equitable, and just society. By guiding decision-making and behavior at all levels, from individual choices to global policies, these ethical principles offer a pathway towards a thriving future for both humanity and the planet.

The potential impacts of widespread adoption of the Earthmate principles (Earthmate Philosophy, Earthmate Science, and Earthmate Ethics) are immense, offering transformative possibilities for how societies interact with the environment and each other. The Earthmate Philosophy, with its emphasis on unity despite diversity, fosters collaboration and empathy, potentially leading to innovative solutions and resilient communities. Earthmate Science provides a comprehensive approach to environmental challenges, potentially unlocking breakthroughs in sustainable technologies and fostering global cooperation. Meanwhile, Earthmate Ethics guides ethical decision-making, promoting sustainability, equity, and justice, potentially leading to more conscientious consumption patterns and greater social cohesion. However, it's important to note that these are potential impacts, and further research and implementation will be necessary to fully understand and realize the benefits of widespread adoption of these principles. Nevertheless, the promise of a more sustainable, equitable, and harmonious world for current and future generations is a compelling vision worth pursuing through the adoption of the Earthmate principles framework.

Innovations and Advancements in the Earthmate Principles

Alongside the potential impact of the Earthmate principles in the three domains of the Earthmate Philosophy, the Earthmate Science, and the Earthmate Ethics, these principles manifest concretely in numerous actions and innovations by institutions and individuals, as highlighted extensively in the previous chapter. In addition to the examples discussed earlier, some of the most prominent manifestations of the Earthmate principles can be observed in the following sectors:

Institutions like The Earth Charter and ClientEarth exemplify the Earthmate Philosophy by promoting principles of unity and interconnectedness. The Earth Charter, a global declaration of fundamental ethical principles for building a just, sustainable, and peaceful world, embodies the spirit of the Earthmate Philosophy through its emphasis on respect for nature and the interconnectedness of all life. Similarly, ClientEarth, a legal advocacy organization, uses the law to protect the environment and promote sustainable development, aligning with the Earthmate Philosophy's call for responsible stewardship.

Among individuals, Leonardo DiCaprio and Greta Thunberg are notable champions of the Earthmate Philosophy. DiCaprio, a renowned global celebrity and environmental activist, utilizes his platform to raise awareness about climate change and advocate for sustainable solutions, reflecting the principles of unity and collective responsibility. Greta Thunberg, a student activist and leader of the youth climate movement, embodies the Earthmate Philosophy through her advocacy for urgent climate action and the recognition of interconnectedness in addressing environmental challenges.

In the realm of the Earthmate Science, initiatives like Environmental Humanities and publications such as The Club of Rome's "Limits to Growth" report demonstrate the application of systems thinking and interdisciplinary approaches to environmental issues. Environmental Humanities integrates philosophical, cultural, and historical perspectives to understand human interactions with the environment, promoting a holistic view aligned with the Earthmate Science. The Club of Rome's report highlighted the interconnectedness of economic growth and environmental sustainability, pioneering the discourse on planetary boundaries.

Pioneering researchers like Rachel Carson, Dr. Jane Goodall, and Dr. James Hansen have significantly advanced our understanding of environmental science. Carson's "Silent Spring" exposed the dangers of pesticides, inspiring the modern environmental movement. Goodall's research on chimpanzees emphasized biodiversity conservation and interconnectedness in ecosystems. Hansen's work on climate modeling has informed climate policy and underscored the urgency of addressing global warming.

Innovation in renewable energy, exemplified by companies like Tesla and SolarCity, showcases practical applications of the Earthmate Science. Their advancements in electric vehicles and solar technologies contribute to reducing carbon emissions and promoting sustainable energy solutions. Additionally, the development of offshore wind farms by companies like Ørsted demonstrates scalable renewable energy infrastructure aligned with the Earthmate Science principles.

In the realm of the Earthmate Ethics, influential figures like Pope Francis and Wangari Maathai have advocated for ethical principles of environmental stewardship. Pope Francis's encyclical "Laudato si'" emphasizes the moral imperative of caring for our common home and promoting integral ecology. Wangari Maathai's Green Belt Movement empowered communities through tree planting and environmental conservation, highlighting the intersection of environmental and social justice.

Documentaries by David Attenborough and advocacy efforts by Bill McKibben raise awareness about environmental ethics and the need for collective action. Attenborough's documentaries showcase the beauty of nature and the urgency of conservation, fostering empathy and ethical consideration for the environment. McKibben's activism underscores the ethical dimensions of climate change and the importance of grassroots movements in driving change.

The Standing Rock Sioux Tribe's resistance against the Dakota Access Pipeline in the United States and local protests against the Miankaleh petrochemical complex in Iran exemplify grassroots efforts for environmental and social justice. These movements highlight the importance of indigenous rights, community empowerment, and ethical considerations in environmental decision-making.

The examples presented across the Earthmate Philosophy, Science, and Ethics underscore the diverse and impactful manifestations of the Earthmate principles in action. From legal advocacy and interdisciplinary research to grassroots activism and celebrity engagement, these initiatives demonstrate a global commitment to unity, interconnectedness, responsibility, and justice in addressing environmental challenges. These instances represent a limited selection, highlighting just a few of the many ways Earthmate ideals can influence and inspire sustainable practices. By embracing the Earthmate principles, individuals and institutions contribute to a more sustainable and harmonious relationship with the planet, paving the way towards a future where ethical considerations guide our actions for the well-being of all life on Earth.

The Role of Principles Earthmate in Shaping the Future

The role of Earthmate principles in shaping the future can be profoundly positive or negligible, contingent upon whether we embrace them widely or disregard them altogether. I examine the transformative role of Earthmate principles in shaping the future of environmental management and sustainability by considering the potential impact of widespread adoption and the innovations and advancements in these principles across nine key levels of application. I also explore how Earthmate principles foster collaboration and shared goals among individuals, communities, and organizations. This analysis includes considering the challenges and obstacles that need to be overcome for effective implementation of Earthmate principles, as well as exploring how these principles can offer solutions to address these challenges.

Ultimately, "The Future of Earthmate" is a call to action. It is a reminder that we have the power to shape the future of our planet and that we must act now to ensure that future generations inherit a world that is healthy, sustainable, and just. By embracing the Earthmate principles, we can create a brighter future for ourselves, for all living things, and for the planet we call home.

The Earthmate principles, encompassing the Earthmate Philosophy, Earthmate Science, and Earthmate Ethics, play a transformative role in shaping the future for individuals across diverse sectors. Beginning with Ordinary People, the adoption of Earthmate principles can empower individuals to recognize their interconnectedness with the environment and promote responsible behaviors, such as sustainable consumption and waste reduction. For Experts & Academia, embracing Earthmate Science fosters interdisciplinary collaboration and innovative research to address complex environmental challenges through systems thinking and sustainable development principles.

Celebrities, Influencers, and Public Figures have a significant platform to amplify Earthmate principles, driving widespread awareness and advocacy for sustainability initiatives. Influential figures like Leonardo DiCaprio leverage their influence to promote environmental stewardship and collective action. Religious and Ethical Leaders integrate Earthmate Ethics into moral teachings, emphasizing responsibility towards nature and advocating for ethical behavior aligned with principles of unity and interdependence.

Environmental reporters, journalists, and communicators play a crucial role in disseminating Earthmate principles to wider audiences, shaping public perceptions and policy discourse. Environmental Activists and Advocates, including youth leaders like Greta Thunberg, mobilize communities and galvanize global movements for climate and environmental justice, reflecting Earthmate's call for collaboration and shared goals.

Environmental Justice and Climate Justice Advocates prioritize equity and fairness in environmental decision-making, aligning with Earthmate Ethics principles of sustainability, responsibility, and justice. Traditional Knowledge Holders preserve indigenous wisdom and practices that harmonize with Earthmate Philosophy's respect for nature and interconnectedness.

Environmental Writers, Filmmakers, and Documentarians amplify Earthmate principles through storytelling and artistic expression, fostering empathy and understanding for environmental challenges and solutions. Their work engages diverse audiences and catalyzes behavioral change.

The potential impact of widespread adoption of Earthmate principles hinges on addressing challenges and obstacles, such as policy inertia, economic interests, and societal behaviors. However, Earthmate principles offer solutions by promoting collaboration, tolerance, and shared goals among individuals, communities, and organizations. By navigating complexities through interdisciplinary approaches and embracing ethical values, Earthmate envisions a harmonious and sustainable future where unity prevails amid the plurality of existence. The integration of Earthmate Philosophy, Science, and Ethics fosters a self-regulating, interconnected whole that prioritizes the well-being of the planet and its inhabitants for current and future generations.

The role of Earthmate principles in shaping the future for Educational and Consultative Institutions is pivotal in fostering a paradigm shift towards environmental management and sustainability. Schools and Environmental Education Programs, alongside Colleges and Universities with Environmental Education Programs, play a fundamental role in instilling Earthmate principles from an early age, cultivating environmental awareness and responsibility. Informal online or offline Environmental Education platforms expand outreach and accessibility, reaching diverse audiences globally. Partnerships and Collaborations targeting environmental education enhance collective efforts, leveraging resources and expertise to advance sustainability agendas. Environmental Consultants and Advisory Services integrate Earthmate principles into strategic planning and decision-making, offering solutions aligned with responsible stewardship and circular economy principles.

Think Tanks and Policy Institutes drive policy innovation by advocating for Earthmate principles within governmental frameworks. R&D Institutions leverage scientific advancements rooted in Earthmate Science, embracing interdisciplinary approaches to address complex environmental challenges. Widespread adoption of Earthmate principles in these institutions fosters collaboration and shared goals among individuals, communities, and organizations, catalyzing collective action towards sustainability. Challenges such as policy inertia and economic constraints require strategic solutions informed by Earthmate Ethics, emphasizing interdependence, responsibility, and equity. Overcoming these obstacles involves fostering global cooperation, bridging understanding gaps, and addressing cultural sensitivities integral to Earthmate's holistic approach. Earthmate principles offer solutions by promoting a delicate balance between economic prosperity, social well-being, and environmental protection, envisioning a sustainable future where humanity collectively commits to responsible stewardship and ecological balance. Through collaborative efforts guided by Earthmate principles, Educational and Consultative Institutions can lead transformative change towards a harmonious relationship between societies and the planet.

The role of Earthmate principles in shaping the future for Media and Culture is profound, influencing the narratives, perspectives, and actions surrounding environmental management and sustainability. Media and Communication platforms serve as crucial channels for disseminating Earthmate principles, highlighting interconnectedness, responsible stewardship, and sustainability. Social Media and Online Activism amplify environmental awareness and advocacy, mobilizing global communities towards shared environmental goals. Earthmate principles intersect with Religious and Ethical Perspectives, emphasizing respect for nature, interdependence, and justice, fostering a deeper moral commitment towards environmental stewardship. Art and Literature embody Earthmate principles through creative expressions, inspiring empathy and action for environmental sustainability.

The transformative impact of Earthmate principles in Media and Culture hinges on widespread adoption and innovative practices that promote collaboration and shared goals. Challenges such as misinformation, apathy, and resistance to change must be addressed to effectively implement Earthmate principles. Earthmate principles offer solutions by fostering interdisciplinary collaborations, amplifying diverse voices, and cultivating a collective sense of responsibility towards the planet. Through strategic engagement with Media and Culture, Earthmate principles catalyze transformative change, empowering individuals, communities, and organizations to envision and realize a harmonious, sustainable future.

The role of Earthmate principles in shaping the future for Profit-Oriented Entities is transformative, offering a pathway towards environmental management and sustainability in various sectors. Earthmate Philosophy, Science, and Ethics provide a holistic framework that can guide Financial Institutions, Industrial Producers, the Mining Industry, and Agriculture towards responsible practices. These principles advocate for sustainable agriculture, agroforestry, and permaculture initiatives, fostering regenerative approaches to production. Furthermore, Earthmate principles drive technological innovations in environmental startups, influencing consumer behavior and supply chains towards sustainability. In the Tourism Industry, Earthmate promotes sustainable tourism and ecotourism practices that respect local environments and communities. In construction and packaging sectors, Earthmate encourages green building, sustainable packaging, and waste management practices, reducing environmental impacts. Public–Private Partnerships are essential in implementing Earthmate principles, fostering collaboration among stakeholders for shared environmental goals. Challenges such as resource constraints and regulatory barriers require innovative solutions guided by Earthmate Ethics, emphasizing responsibility, interdependence, and equity. Overcoming these obstacles will require a collective effort to align profit-oriented entities with Earthmate principles, ensuring a sustainable and harmonious future for all stakeholders and the planet.

The Earthmate principles play a pivotal role in shaping the future for Non-Profit Entities dedicated to environmental and social causes. Environmental Justice & Climate Justice Associations, Wildlife Conservation Associations, Climate Change Mitigation and Adaptation Associations,

Ocean Protection Associations, Community-Based Conservation Initiatives, Green Philanthropy and Impact Investing, Legal Advocacy and Environmental Law NGOs, and Crisis Response and Disaster Relief NGOs are increasingly adopting Earthmate principles to drive transformative change. Widespread adoption of Earthmate Philosophy, Science, and Ethics fosters collaboration and shared goals among these entities, facilitating interdisciplinary approaches and global cooperation to address environmental challenges. By embracing Earthmate principles, non-profit organizations can navigate obstacles such as limited resources, policy constraints, and competing priorities, leveraging innovative solutions grounded in sustainability, responsibility, and justice. Earthmate principles offer a roadmap for effective environmental management and sustainability by promoting equitable partnerships, holistic strategies, and collective action toward a harmonious future for current and future generations.

The Earthmate principles play a crucial role in shaping the future for Governmental Entities engaged in environmental management and sustainability. Legislative bodies, Legal and Regulatory Frameworks, Law Enforcement Entities, the Judiciary, Governmental Organizations for Wildlife Conservation, Government Water Resource Management Entities, Governmental Institutions for Meteorology, Climate Science, and Weather Forecasting, Environmental Government Agencies, and Urban Planning, Green Infrastructure, and Sustainable Development are key stakeholders poised to benefit from adopting Earthmate principles. Widespread adoption of Earthmate Philosophy, Science, and Ethics can drive transformative change by promoting collaborative governance, evidence-based policymaking, and integrated planning approaches. These principles foster interdisciplinary collaboration, aligning policies with scientific insights and ethical considerations. Challenges such as policy inertia, resource constraints, and competing interests can be addressed through innovative solutions grounded in sustainability, responsibility, and justice. By embracing Earthmate principles, Governmental Entities can enhance resilience, promote environmental stewardship, and advance sustainable development goals, ultimately contributing to a harmonious future for communities and ecosystems alike.

The integration of Earthmate principles holds immense promise in shaping the future of Health and Environment, encompassing the Healthcare Industry, Medical Infrastructure and Health Systems, Environmental Psychology, Community-Based Health Initiatives, and Global Health and Environmental Organizations. Through the adoption of Earthmate Philosophy, Science, and Ethics, transformative changes can be catalyzed, leading to a more sustainable and resilient health ecosystem. By recognizing the interconnectedness between human health and the environment, Earthmate principles encourage collaborative efforts to address environmental challenges that impact public health. Innovations in medical infrastructure and health systems can incorporate sustainable practices, reducing their environmental footprint and promoting community well-being. Environmental psychology studies can inform policies and interventions aimed at fostering pro-environmental behaviors and promoting mental health. Community-based health initiatives can leverage Earthmate principles to empower local communities in environmental stewardship, enhancing both environmental and human health outcomes. Furthermore, global health and environmental organizations can play a pivotal role in advocating for policies that prioritize sustainability and equity, fostering collaboration among nations to address global environmental challenges. Challenges such as resource constraints, policy inertia, and lack of awareness need to be addressed through education, advocacy, and policy interventions. Earthmate principles offer a comprehensive framework for addressing these challenges, promoting a harmonious relationship between health and the environment for the well-being of current and future generations.

Earthmate principles play a crucial role in shaping the future for international organizations by providing a comprehensive framework for addressing global environmental challenges. United Nations bodies and specialized agencies, alongside global conservation and biodiversity organizations, can be increasingly embracing Earthmate Philosophy, Science, and Ethics to foster collaboration and shared goals among nations, communities, and organizations worldwide. By recognizing the interconnectedness of all living things and advocating for responsible stewardship of the planet, Earthmate principles drive transformative change in international environmental agreements, protocols, conventions, and treaties. These principles promote the alignment of global environmental funds and financing mechanisms with sustainable development objectives, fostering innovative approaches to address complex environmental issues. However, challenges such as political obstacles, resource constraints, and differing priorities among member states must be overcome for effective implementation of Earthmate principles. Solutions lie in promoting dialogue, cooperation, and capacity-building initiatives to enhance global environmental governance and facilitate international research and scientific collaboration. Through widespread adoption of Earthmate principles, international organizations can play a pivotal role in shaping a harmonious and sustainable future for current and future generations, ensuring the well-being of the planet and its inhabitants on a global scale.

Earthmate principles play a pivotal role in shaping the future of special environment-oriented entities by providing a comprehensive framework that promotes sustainability, responsible stewardship, and ethical practices. These entities include standardization and certification bodies, carbon offset and environmental credit providers, biodiversity offsetting initiatives, innovations in carbon markets and offset financing, and PES programs. By integrating Earthmate Philosophy, Science, and Ethics, these entities can drive transformative change in environmental management and sustainability.

The Earthmate Philosophy, rooted in the concept of UDP, emphasizes the interconnectedness of nature and humanity. This philosophy can guide special environment-oriented entities to adopt holistic approaches that consider the broader ecological context and the interdependence of ecosystems. By embracing UDP, these entities can foster collaboration among stakeholders and promote shared goals that prioritize environmental well-being alongside economic and social interests.

Earthmate Science provides a robust foundation for innovative practices within these entities. Aligned with systems thinking and sustainable development principles, Earthmate Science encourages a nuanced understanding of complex environmental issues. This scientific approach supports the development of innovative technologies and methodologies for carbon offsetting, biodiversity conservation, and ecosystem restoration.

Ethical considerations guided by Earthmate Ethics are essential for special environment-oriented entities. These principles emphasize responsibility, equity, and justice in environmental decision-making. By promoting ethical practices, such as transparent certification processes and equitable distribution of benefits from ecosystem services, Earthmate Ethics can enhance trust and collaboration among stakeholders.

Widespread adoption of Earthmate principles within these entities can have transformative impacts on environmental management and sustainability. It can drive innovation in carbon markets, biodiversity conservation, and ecosystem services valuation, leading to more effective and equitable solutions to environmental challenges. Furthermore, Earthmate principles foster collaboration by promoting inclusive decision-making processes, fostering partnerships, and addressing cultural sensitivities.

Challenges and obstacles to effective implementation of Earthmate principles include regulatory complexities, market volatility, and capacity constraints. Overcoming these challenges requires coordinated efforts to build institutional capacity, enhance policy frameworks, and promote international cooperation. Earthmate principles offer solutions by advocating for adaptive governance, inclusive stakeholder engagement, and innovative financing mechanisms to support sustainable practices. In conclusion, Earthmate principles have the potential to revolutionize the practices of special environment-oriented entities by promoting holistic, science-based, and ethical approaches to environmental management and sustainability. By embracing these principles, these entities can contribute significantly to a more resilient, equitable, and sustainable future for our planet.

Ultimately, the role of Earthmate principles in shaping the future is profound and hinges on widespread adoption across diverse sectors. By integrating Earthmate Philosophy, Science, and Ethics, transformative changes can be catalyzed in environmental management and sustainability practices. The potential impact of these principles on collaboration, innovation, and collective action is substantial, fostering shared goals among individuals, communities, and organizations worldwide.

The future of Earthmate is a call to action—a reminder that we possess the ability to shape the destiny of our planet and ensure a healthy, sustainable, and just world for future generations. Embracing Earthmate principles empowers us to create a brighter future for all living beings and the planet we call home.

Through this comprehensive analysis across various levels of application—from ordinary people to governmental entities, educational institutions to profit-oriented entities, and beyond—we see the transformative potential of Earthmate principles. They offer a roadmap for responsible stewardship, collaboration, and ethical decision-making, aligning societal actions with the interconnectedness of all life on Earth.

However, challenges such as policy inertia, economic constraints, and societal behaviors must be addressed for effective implementation of Earthmate principles. Solutions lie in promoting interdisciplinary collaboration, fostering global cooperation, and advocating for responsible practices guided by Earthmate Ethics.

In summary, the integration of Earthmate principles fosters a self-regulating, interconnected whole that prioritizes the well-being of the planet and its inhabitants. I hold the belief that embracing effective globalism will encounter significant economic, cultural, and political resistance; however, it is challenging to envision sustainable planetary futures without advancing in that direction. By embracing these principles, we can collectively navigate the complexities of our time and work towards a harmonious and sustainable future—one that ensures the flourishing of ecosystems, societies, and future generations.

The manufacturer's authorised representative in the EU is Springer Nature Customer Service Centre GmbH, Europaplatz 3, 69115 Heidelberg, Germany. If you have any concerns regarding our products, please contact ProductSafety@springernature.com

Printed and bound by CPI Group (UK) Ltd, Croydon, CR0 4YY
23/03/2026
02076446-0001